The Mechanics of Fracture and Fatigue

AN INTRODUCTION

The Mechanics of Fracture and Fatigue

AN INTRODUCTION

A. P. PARKER
B.Sc., Ph.D., C.Eng.
Royal Military College of Science, Shrivenham

LONDON   NEW YORK
E. & F. N. SPON LTD

First published 1981
by E. & F. N. Spon Ltd
11 New Fetter Lane, London EC4P 4EE

Published in the USA
by E. & F. N. Spon
in association with Methuen, Inc.
733 Third Avenue, New York NY 10017

ISBN 0 419 11460 2 (cased)
ISBN 0 419 11470 X (paperback)

Printed in the United States of America

© 1981 A. P. Parker

This title is available in both hardbound and paperback editions. The paperback edition is sold subject to the condition that it shall not, by way of trade or otherwise, be lent, re-sold, hired out, or otherwise circulated without the publisher's prior consent in any form of binding or cover other than that in which it is published and without a similar condition including this condition being imposed on the subsequent purchaser.

All rights reserved. No part of this book may be reprinted, or reproduced or utilized in any form or by any electronic, mechanical or other means, now known or hereafter invented, including photocopying and recording, or in any information storage and retrieval system, without permission in writing from the Publisher.

British Library Cataloguing in Publication Data

Parker, A P
    The mechanics of fracture and fatigue.
    1. Materials — Fatigue   2. Fracture mechanics
    I. Title
    624'.176   TA409   80-41678

ISBN 0-419-11460-2
ISBN 0-419-11470-X Pbk

Contents

Preface		*page* ix
1 Introductory Theory of Elasticity		1
1.1	Introduction	1
1.2	Stress	1
	1.2.1 *Equations of equilibrium*	4
	1.2.2 *Stress transformations*	5
1.3	Strain	6
	1.3.1 *Compatibility of strains*	8
1.4	Physical law	8
1.5	Plane deformation	8
1.6	Airy stress deformation	10
*1.7	Complex stress functions	11
*1.8	Conformal mapping	15
	*1.8.1 *Elliptical cutout in infinite sheet under uniaxial tension*	17
References		20
2 Energy Considerations		21
2.1	Introduction	21
2.2	General	21
2.3	The Griffith Crack	21
	2.3.1 *Potential energy*	22
	2.3.2 *Surface energy*	24
	2.3.3 *Total energy*	24
	2.3.4 *Energy release rates*	24
2.4	Compliance	25
	2.4.1 *Loading conditions*	26
References		27
3 Stresses and Displacements in Cracked Bodies		28
3.1	Introduction	28
3.2	Stresses in cracked bodies	28

* Readers wishing to avoid the more analytical sections of the book may omit those which are asterisked.

3.3	Stress intensity factor	30
3.4	Fracture toughness	31
3.5	Crack shape	33
3.6	Energy released	34
3.7	Modes of crack tip deformation	35
*3.8	Westergaard stress function	39
3.9	The configuration correction factor	44
*3.10	The Williams stress function	44
	Problems	46
	References	48

4 Determination of Stress Intensity Factors 49

4.1	Introduction	49
4.2	Analytical	49
4.3	Green's functions	51
4.4	Weight function techniques	51
	4.4.1 *Derivation*	52
	4.4.2 *Additional remarks*	62
*4.5	Boundary collocation	63
	*4.5.1 *Collocation of the complex stress functions*	64
	*4.5.2 *Symmetry and anti-symmetry properties*	65
	*4.5.3 *Special cases of the geometry*	66
	*4.5.4 *Outline programming technique*	67
	*4.5.5 *Mapping–collocation techniques*	68
*4.6	Finite element methods	70
	*4.6.1 *Non-singular crack tip representations*	70
	*4.6.2 *Singular elements*	72
*4.7	Integral equations	73
	*4.7.1 *Formulation*	74
*4.8	Boundary methods	74
4.9	The compounding method	76
4.10	Experimental methods	78
	Appendix A: Uniqueness of the weight function	79
	Appendix B: The weight function for mixed boundary condition problems	81
	References	84

5 Mixed-mode Fracture Mechanics 89

5.1	Introduction	89
5.2	The effective stress intensity factor in mixed mode	89
5.3	Crack direction in mixed mode	91

*5.4	The strain energy density criterion	93
	*5.4.1 *The strain energy density field for plane problems*	93
	*5.4.2 *Physical significance of the strain energy density concept*	95
*5.5	Crack path stability	97
	References	99

6 Crack Tip Plasticity and Associated Effects — 101

6.1	Introduction	101
6.2	Irwin's plastic zone model	101
6.3	Dugdale's plastic zone model	103
6.4	Plastic zone shapes	106
*6.5	Types of failures	108
*6.6	R−curves	111
*6.7	Plane strain fracture toughness testing	113
*6.8	Failure criteria in the presence of moderate plasticity	117
	*6.8.1 *Crack tip opening displacement (COD)*	117
	*6.8.2 *The J integral*	118
*6.9	A note on general yield loads	120
*6.10	The failure assessment diagram	120
	References	121

7 Fatigue Crack Growth — 123

7.1	Introduction	123
7.2	Stress intensity factor range	123
7.3	Empirical crack growth rate results	124
7.4	Use of crack growth law	127
7.5	Other factors affecting fatigue crack growth rate	132
*7.6	Variable-amplitude loading	133
*7.7	The rainflow method	134
*7.8	Mixed-mode loading	135
	References	136

8 The Fracture Mechanics Design Process — 138

8.1	Introduction	138
8.2	Crack detection techniques	138
8.3	Initial flaw sizes	139
8.4	Fail-safe and safe-life design concepts	140
8.5	Locating stress intensity factor solutions	141
8.6	Locating fracture toughness and fatigue crack growth data	142
8.7	Design examples	142
	References	160
	Index	163

Preface

A problem of considerable, and increasing importance within the fields of civil, mechanical, aeronautical, marine and military engineering is the predominantly brittle failure of structures. This type of failure has been observed to occur under both constant and cyclic loading conditions. *Fracture mechanics* has evolved as a result of attempts to understand and prevent such failures.

Historically, fracture mechanics developed along parallel lines at the microscopic and macroscopic levels. The latter, in the form of solid mechanics, produced numerous, somewhat daunting contributions in the mathematical theory of elasticity as applied to structures containing crack-like defects. This may have inhibited the teaching of fracture mechanics at undergraduate and immediate postgraduate level as an extension of a conventional course on solid mechanics.

The object of this short monograph is simply to assist the teaching of fracture mechanics as a logical extension of an undergraduate course in stress analysis, and to provide a self-teaching aid for the postgraduate or designer.

Chapter 1 is deliberately framed to allow the reader who has completed an undergraduate course in stress analysis to 'read in' through familiar territory, whilst not excluding the individual who omitted stress analysis as a final year option.

In Chapter 2 the important energy considerations are introduced, and the importance of energy release rate is demonstrated, whilst Chapter 3 covers the relationship between the energy release rate and the crack tip stress and displacement field, the different modes of crack tip deformation, and analytical methods which demonstrate the importance of both loading and geometry in determining the crack tip stress field singularity, termed the *stress intensity factor*. This philosophy continues in Chapter 4 wherein methods of determining stress intensity are considered. This chapter may be studied at a level appropriate to the reader, whether the requirement is an understanding of the limitations of a particular solution method, or the development of a particular analysis technique. The reader is recommended to study Section 4.4 relating to *weight function* methods. The weight function is a relatively new tool in fracture mechanics, but possesses considerable potential and flexibility.

Following a study of mixed-mode effects in Chapter 5, the important question of the degree and significance of plasticity effects, including validity of tests, is covered in Chapter 6. In the last two chapters the reader meets the concept of *fatigue crack growth* arising from cyclic loading of a structure, and methods by which fracture mechanics parameters may be used to quantify this phenomenon. Finally, the fail-safe and safe-life *design* concepts are described, and fracture mechanics criteria are applied in some straightforward design exercises.

I wish to extend my thanks to various colleagues who have helped directly or indirectly in the preparation of this monograph, in particular to Mrs Pat Watson who prepared the manuscript.

November 1979

A.P.P.

1 Introductory Theory of Elasticity

1.1 Introduction

In order to predict failure of engineering components it is necessary to understand the concepts of stress and strain, and to link these via a physical law.

Historically, stresses and strains have been predicted for two-dimensional linear elastic problems, by use of stress functions of the Airy type. More recently, complex stress functions have been employed because of certain advantages, in particular the use of conformal mapping techniques in the solution of problems involving complicated boundary shapes.

Each of the above concepts is now examined in turn.

1.2 Stress

Fig. 1.1 shows a three-dimensional body in equilibrium under the action of external forces $F_n(n = 1, 2, 3, \ldots)$. The body is now considered to be 'sliced' at cross section C. The forces which were acting through this cross section are maintained in order to preserve equilibrium. Confining our attention to a small

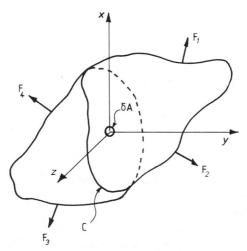

Fig. 1.1 *Three-dimensional body under general loading system.*

area δA of the cut surface, there will be a proportion of the total force at the cross section acting on δA, namely δF.

The force δF may be resolved into components, one perpendicular to the cut surface, δP, and one in the plane of the cut surface, δQ, as shown in Fig. 1.2. The intensity of these forces (i.e. the force per unit area) at a point is termed the stress. There are stresses associated with both the perpendicular and the tangential components of the force δF, and these are given by:

$$\text{direct stress,} \quad \sigma = \lim_{\delta A \to 0} \frac{\delta P}{\delta A} \qquad (1.1)$$

$$\text{shear stress,} \quad \tau = \lim_{\delta A \to 0} \frac{\delta Q}{\delta A}. \qquad (1.2)$$

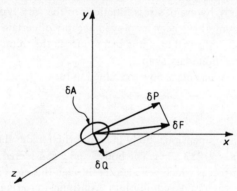

Fig. 1.2 *Normal and tangential force components.*

In general, a rectangular element within a three-dimensional body will have both direct and shear stresses acting on each of six faces. In order to form a coherent reference system it is necessary to resolve the stress on each face in accordance with the coordinate system, as illustrated in Fig. 1.3. This produces nine stress components, namely,

$$\sigma_x \quad \tau_{xy} \quad \tau_{xz}$$
$$\sigma_y \quad \tau_{yx} \quad \tau_{yz}$$
$$\sigma_z \quad \tau_{zx} \quad \tau_{zy}.$$

Note that a single suffix for direct stress indicates the direction of action of the stress. The sign convention is that a positive direct stress tends to increase the dimensions of the element along its line of action, a negative direct stress tends to reduce it. In the case of shear stresses, the first suffix indicates the direction of the normal to the plane on which it acts, the second suffix indicates the direction in which it acts.

INTRODUCTORY THEORY OF ELASTICITY

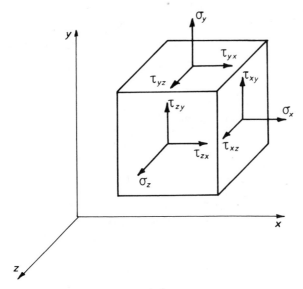

Fig. 1.3 *Stresses on a three-dimensional element.*

In order to preserve moment equilibrium in the element:

$$\tau_{zx} = \tau_{xz}, \quad \tau_{zy} = \tau_{yz}, \quad \tau_{yx} = \tau_{xy}. \tag{1.3}$$

In some cases of configurations which possess geometrical or loading symmetries it may be advantageous to use cylindrical polar coordinates as shown in Fig. 1.4.

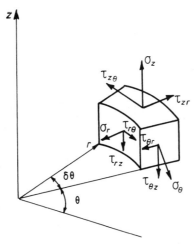

Fig. 1.4 *Cylindrical stress system.*

The equivalent components of stress are given by:

$$\begin{array}{ccc} \sigma_r & \tau_{r\theta} & \tau_{rz} \\ \sigma_\theta & \tau_{\theta r} & \tau_{\theta z} \\ \sigma_z & \tau_{zr} & \tau_{z\theta} \end{array}$$

whilst moment equilibrium is ensured if:

$$\tau_{r\theta} = \tau_{\theta r}, \quad \tau_{\theta z} = \tau_{z\theta}, \quad \tau_{zr} = \tau_{rz}. \tag{1.4}$$

1.2.1 *Equations of equilibrium*

Consider an element having sides of length δx, δy, δz, being subjected to a general stress system in which increments of stress may occur. If we consider only forces in the x-direction, the stresses which must be considered are shown in Fig. 1.5.

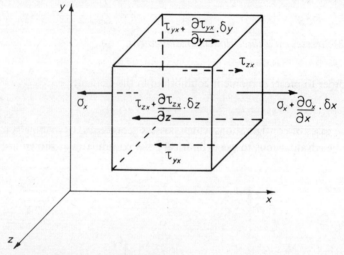

Fig. 1.5 *Stress equilibrium in the x-direction.*

Neglecting body forces, equilibrium in the x-direction is satisfied by:

$$\frac{\partial \sigma_x}{\partial x} + \frac{\partial \tau_{xy}}{\partial y} + \frac{\partial \tau_{xz}}{\partial z} = 0 \tag{1.5}$$

similarly y- and z-direction equilibrium is satisfied by:

$$\frac{\partial \sigma_y}{\partial y} + \frac{\partial \tau_{yz}}{\partial z} + \frac{\partial \tau_{yx}}{\partial x} = 0$$

$$\frac{\partial \sigma_z}{\partial z} + \frac{\partial \tau_{zx}}{\partial x} + \frac{\partial \tau_{zy}}{\partial y} = 0. \tag{1.6}$$

INTRODUCTORY THEORY OF ELASTICITY

1.2.2 Stress transformations

A two-dimensional rectangular element is illustrated in Fig. 1.6. In order to determine normal and shear stress components based on an (x', y') coordinate system, from those in the (x, y) coordinate system, where the new axes are formed by rotating the original axes through a positive angle θ, it is necessary to consider the equilibrium of the element. The resulting equations are well known [1]* namely:

$$\sigma_{x'} = \frac{\sigma_x + \sigma_y}{2} - \left(\frac{\sigma_y - \sigma_x}{2}\right) \cos 2\theta + \tau_{xy} \sin 2\theta$$

$$\sigma_{y'} = \frac{\sigma_x + \sigma_y}{2} + \left(\frac{\sigma_y - \sigma_x}{2}\right) \cos 2\theta - \tau_{xy} \sin 2\theta \qquad (1.7)$$

$$\tau_{x'y'} = \tau_{xy} \cos 2\theta + \left(\frac{\sigma_y - \sigma_x}{2}\right) \sin 2\theta.$$

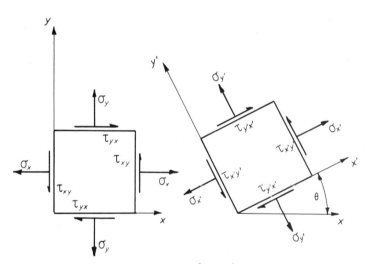

Fig. 1.6 *Stress at a point, coordinate transformation.*

These may be presented in the alternative form [2]:

$$\sigma_{x'} + \sigma_{y'} = \sigma_x + \sigma_y$$

$$\sigma_{y'} - \sigma_{x'} + 2i\tau_{x'y'} = (\sigma_y - \sigma_x + 2i\tau_{xy})e^{2i\theta} \qquad (1.8)$$

where $i = (-1)^{1/2}$. Equations (1.7) or (1.8) form the basis of the Mohr's stress circle construction [2].

---

* References are included at the end of each chapter.

By allowing θ to vary, it is possible to identify two orthogonal planes on which the shear stress is zero, and the direct stress takes either a maximum or a minimum value. These stationary values of direct stress are designated principal stresses, σ_1 and σ_2, and may be obtained from the relationship:

$$\left.\begin{matrix}\sigma_1\\ \sigma_2\end{matrix}\right\} = \left(\frac{\sigma_x + \sigma_y}{2}\right) \pm \left[\left(\frac{\sigma_x - \sigma_y}{2}\right)^2 + \tau_{xy}^2\right]^{1/2}.$$

In general, of course, for a three-dimensional problem there will be three principal stresses, σ_1, σ_2, σ_3. It is common practice to adopt the convention $\sigma_1 \geqslant \sigma_2 \geqslant \sigma_3$, and in plane problems it is necessary to compare through-the-thickness stresses before putting the principal stresses in order. This is of particular importance when yield criteria, which are normally expressed in terms of principal stresses, are being evaluated, as in Chapter 6.

1.3 Strain

A small rectangular element ABCD is illustrated in Fig. 1.7. After straining, changes of length and shape have occurred, and the element occupies A'B'C'D'. In a manner analogous to the stresses, the strains are of two kinds: direct strain, ϵ_i (a measure of the proportionate increase in length), and shear strain, ϵ_{ij} (a measure of the proportionate distortion of the corners of the element).

From Fig. 1.7 the direct strain in the x-direction, ϵ_x defined as the increase in

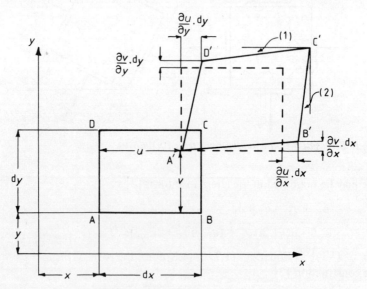

Fig. 1.7 *Strain in a two-dimensional element.*

INTRODUCTORY THEORY OF ELASTICITY

length/original length, becomes:

$$\epsilon_x = \frac{A'B' - AB}{AB}$$

neglecting terms of order $(\partial u/\partial x)^2$:

$$\epsilon_x = \frac{(\partial u/\partial x)\delta x}{\delta x} = \frac{\partial u}{\partial x}.$$

The complete set of equations for direct strains is:

$$\epsilon_x = \frac{\partial u}{\partial x}, \quad \epsilon_y = \frac{\partial v}{\partial y}, \quad \epsilon_z = \frac{\partial w}{\partial z} \tag{1.9}$$

where u, v, w are displacements in x-, y-, z-directions respectively. The single suffix indicates the direction of strain, positive strains are associated with an increase in length, negative with a decrease in length.

In order to obtain a measure of the distortion, we note that

$$\text{slope (1)} = \frac{\text{increase in height}}{\text{horizontal length}} = \frac{\partial v}{\partial x} = \frac{\text{angle in radians}}{\text{(small angles)}}$$

$$\text{slope (2)} = \frac{\text{increase in height}}{\text{horizontal length}} = \frac{\partial u}{\partial y} = \frac{\text{angle in radians}}{\text{(small angles)}}.$$

Shear strain, ϵ_{xy}, is defined as half the decrease in angle of the outermost corner of the element, thus:

$$\epsilon_{xy} = \frac{1}{2}\left(\frac{\partial v}{\partial x} + \frac{\partial u}{\partial y}\right).$$

(Note: This is the so-called 'mathematical' definition of shear strain. The 'engineering' shear strain, γ_{xy}, is defined by $\gamma_{xy} = 2\epsilon_{xy}$.)

The double suffix notation is analogous to that used for shear stresses. A positive shear stress is one which tends to reduce the angle of the corner of the element which is outermost with respect to positively directed coordinate axes.

The complete set of equations relating strains to displacements is:

$$\epsilon_{xy} = \frac{1}{2}\left(\frac{\partial v}{\partial x} + \frac{\partial u}{\partial y}\right)$$

$$\epsilon_{xz} = \frac{1}{2}\left(\frac{\partial w}{\partial x} + \frac{\partial u}{\partial z}\right) \tag{1.10}$$

$$\epsilon_{yz} = \frac{1}{2}\left(\frac{\partial w}{\partial y} + \frac{\partial v}{\partial x}\right).$$

Restricting our attention to the (x, y) plane we are left with the following equations:

$$\epsilon_x = \frac{\partial u}{\partial x}, \quad \epsilon_y = \frac{\partial v}{\partial y}, \quad \epsilon_{xy} = \frac{1}{2}\left(\frac{\partial v}{\partial x} + \frac{\partial u}{\partial y}\right). \tag{1.11}$$

1.3.1 Compatibility of strains

The three components of strain are derived from two components of displacement so that some restraints must be placed on 'allowable' strains. The strains must be compatible. By differentiating equations (1.11), we obtain:

$$\frac{\partial^2 \epsilon_x}{\partial y^2} = \frac{\partial^3 u}{\partial x \partial y^2}, \quad \frac{\partial^2 \epsilon_y}{\partial x^2} = \frac{\partial^3 v}{\partial x^2 \partial y}, \quad 2\frac{\partial^2 \epsilon_{xy}}{\partial x \partial y} = \frac{\partial^3 u}{\partial x \partial y^2} + \frac{\partial^3 v}{\partial y \partial x^2} \tag{1.12}$$

thus

$$\frac{\partial^2 \epsilon_x}{\partial y^2} + \frac{\partial^2 \epsilon_y}{\partial x^2} - 2\frac{\partial^2 \epsilon_{xy}}{\partial x \partial y} = 0. \tag{1.13}$$

Equation (1.13) is the compatibility requirement expressed in terms of strains.

1.4 Physical Law

The relationship between stress and strain for an elastic material, termed Hooke's law, is well known [1], and for the three-dimensional case is given by:

$$\epsilon_x = \frac{1}{E}[\sigma_x - v(\sigma_y + \sigma_z)], \quad \epsilon_{yz} = \frac{1+v}{E}\tau_{yz},$$

$$\epsilon_y = \frac{1}{E}[\sigma_y - v(\sigma_z + \sigma_x)], \quad \epsilon_{zx} = \frac{1+v}{E}\tau_{zx}, \tag{1.14}$$

$$\epsilon_z = \frac{1}{E}[\sigma_z - v(\sigma_x + \sigma_y)], \quad \epsilon_{xy} = \frac{1+v}{E}\tau_{xy},$$

where E is Young's modulus (or the modulus of elasticity), and v is Poisson's ratio.

1.5 Plane deformation

The problem of finding stresses and displacements at points within a loaded body is considerably simplified if it can be assumed, either that there is no change in the distribution of stress, or of strain over the (x, y) plane, in the z-direction.

INTRODUCTORY THEORY OF ELASTICITY

Two cases of this plane deformation are:

(a) *Plane stress*. This is the state of stress which may be assumed to exist in a thin sheet, which is considered incapable of supporting stresses through the thickness, i.e. in the z-direction. Thus any stress having a z suffix may be set to zero, yielding:

$$\sigma_z = \tau_{zx} = \tau_{zy} = 0.$$

This reduces the stress–strain relationships (equation (1.14)) to:

$$\epsilon_x = \frac{1}{E}(\sigma_x - v\sigma_y)$$

$$\epsilon_y = \frac{1}{E}(\sigma_y - v\sigma_x) \quad (1.15)$$

$$\epsilon_{xy} = \frac{1+v}{E}\tau_{xy}.$$

(b) *Plane strain*. This models a plate which is sufficiently thick to prevent through-the-thickness strains. Thus any strain having a z suffix may be set to zero, giving:

$$\epsilon_z = \epsilon_{zx} = \epsilon_{zy} = 0.$$

This reduces the stress–strain relationships, after some manipulation, to:

$$\epsilon_x = \frac{1-v^2}{E}\left(\sigma_x - \frac{v}{1-v}\sigma_y\right)$$

$$\epsilon_y = \frac{1-v^2}{E}\left(\sigma_y - \frac{v}{1-v}\sigma_x\right) \quad (1.16)$$

$$\epsilon_{xy} = \frac{1-v^2}{E}\left(1 + \frac{v}{1-v}\tau_{xy}\right).$$

These equations are of exactly the same form as those for plane stress (equation (1.15)) if E is replaced by $E/(1-v^2)$ and v is replaced by $v/(1-v)$. We shall continue our analysis for plane stress only, recognizing that the modification to plane strain is straightforward.

Rewriting the strain compatibility conditions (equation (1.13)) in terms of stresses by substituting from the plane stress equations (1.15), we obtain:

$$\frac{\partial^2}{\partial y^2}(\sigma_x - v\sigma_y) + \frac{\partial^2}{\partial x^2}(\sigma_y - v\sigma_x) + 2(1+v)\frac{\partial^2}{\partial x \partial y}\tau_{xy} = 0. \quad (1.17)$$

Equation (1.17) is the compatibility requirement expressed in terms of stresses.

Any combination of stresses and strains within a body must satisfy these requirements, and also those of equilibrium. The equations of equilibrium (equations (1.5), (1.6)) are reduced under plane stress conditions in the absence of body forces to:

$$\frac{\partial \sigma_x}{\partial x} + \frac{\partial \tau_{xy}}{\partial y} = 0$$

$$\frac{\partial \sigma_y}{\partial y} + \frac{\partial \tau_{xy}}{\partial x} = 0. \tag{1.18}$$

1.6 Airy stress function

Consider a function of x and y, $F(x, y)$, normally termed the Airy stress function, from which stresses may be derived as follows:

$$\sigma_x = \frac{\partial^2 F}{\partial y^2}, \quad \sigma_y = \frac{\partial^2 F}{\partial x^2}, \quad \tau_{xy} = -\frac{\partial^2 F}{\partial x \partial y}. \tag{1.19}$$

It is a straightforward procedure to confirm by substituting from equations (1.19) into equations (1.18) that the Airy stresses satisfy the equilibrium conditions. By inserting the stresses from equations (1.19) into the compatibility equation (1.17) we obtain:

$$\frac{\partial^4 F}{\partial x^4} + 2\frac{\partial^4 F}{\partial x^2 \partial y^2} + \frac{\partial^4 F}{\partial y^4} = 0 \tag{1.20}$$

which may also be written:

$$\left(\frac{\partial^2}{\partial x^2} + \frac{\partial^2}{\partial y^2}\right)\left(\frac{\partial^2}{\partial x^2} + \frac{\partial^2}{\partial y^2}\right) F = 0 \tag{1.21}$$

or in the more compact form:

$$\nabla^4 F = 0 \tag{1.22}$$

where ∇^2 is the harmonic operator: $(\partial^2/\partial x^2) + (\partial^2/\partial y^2)$.

Numerous stress functions of the Airy type are available, together with their associated boundary conditions [1]. Certain such functions which can be expressed concisely in terms of polar coordinates are associated with the name of Boussinesq [3]. Nevertheless, considerable flexibility will be gained by moving on directly to consider complex stress functions. Such functions may be considered as a general case of the Airy and Boussinesq functions. They possess certain advantages:

(a) They are generally more compact.
(b) They may be derived directly from boundary conditions in certain cases.

INTRODUCTORY THEORY OF ELASTICITY

(c) They permit the use of conformal mapping techniques, of particular value for cutouts and cracks.

(d) They permit the use of sophisticated numerical techniques, discussed in Chapter 4.

1.7 Complex stress functions

Whilst the Airy stress function is an admissible function it is not necessarily unique in this respect. Consider now stresses defined by two functions, the first being the Airy function $F(x, y)$, the second some other function $B(x, y)$, such that:

$$\sigma_x = \frac{\partial^2 F}{\partial x^2} - 2\frac{\partial^2 B}{\partial x \partial y}$$

$$\sigma_y = \frac{\partial^2 F}{\partial y^2} + 2\frac{\partial^2 B}{\partial x \partial y} \qquad (1.23)$$

$$\tau_{xy} = \frac{\partial^2 F}{\partial x \partial y} - \frac{\partial^2 B}{\partial y^2} + \frac{\partial^2 B}{\partial x^2}.$$

By substitution from these equations into equations (1.19) and (1.17), we see that both equilibrium and compatibility are satisfied provided:

$$\frac{\partial}{\partial x}(\nabla^2 F) - \frac{\partial}{\partial y}(\nabla^2 B) = 0$$

$$\frac{\partial}{\partial y}(\nabla^2 F) + \frac{\partial}{\partial x}(\nabla^2 B) = 0. \qquad (1.24)$$

These are the Cauchy–Riemann conditions which ensure that $(\nabla^2 F + i\nabla^2 B)$ is an analytic function. Equations (1.24) may be solved by introducing a particular combination of analytic functions $\phi(z)$ and $\psi(z)$ [4] such that:

$$F + iB = \bar{z}\phi(z) + \int \psi(z)\,dz \qquad (1.25)$$

where $z = x + iy$, $\bar{z} = x - iy$. This solution may be verified by noting that, on differentiation

$$\left(\frac{\partial^2}{\partial x^2} + \frac{\partial^2}{\partial y^2}\right)(F + iB) = 4\phi'(z)$$

and since any derivative of an analytic function is itself an analytic function, this gives a solution which satisfies equations (1.24). (Note that the Airy function, $F = \text{Re}\{\bar{z}\phi(z) + \int\psi(z)\,dz\}$, and that it is always possible to obtain the equivalent Airy function by this method.)

Noting that

$$F = \text{Re}\left\{\bar{z}\phi(z) + \int \psi(z)\,dz\right\}$$
$$B = \text{Im}\left\{\bar{z}\phi(z) + \int \psi(z)\,dz\right\} \quad (1.26)$$

where Re and Im are real and imaginary parts respectively, it is now possible to write expressions for the stresses in terms of $\phi(z)$ and $\psi(z)$:

$$\sigma_x = 2\,\text{Re}\{\phi'(z)\} - \text{Re}\{\bar{z}\phi''(z) + \psi'(z)\}$$
$$\sigma_y = 2\,\text{Re}\{\phi'(z)\} + \text{Re}\{\bar{z}\phi''(z) + \psi'(z)\} \quad (1.27)$$
$$\tau_{xy} = \text{Im}\{\bar{z}\phi''(z) + \psi'(z)\}.$$

Combinations of the above stresses are:

$$\sigma_x + \sigma_y = 4\,\text{Re}\{\phi'(z)\}$$
$$\sigma_y - \sigma_x + 2i\tau_{xy} = 2[\bar{z}\phi''(z) + \psi'(z)]. \quad (1.28)$$

which are directly comparable with the stress combinations of equation (1.8), which define the transformation requirements. Thus we may immediately write an expression for change of axes in the form:

$$\sigma_{x'} + \sigma_{y'} = 4\,\text{Re}\{\phi'(z)\}$$
$$\sigma_{y'} - \sigma_{x'} + 2i\tau_{x'y'} = 2[\bar{z}\phi''(z) + \psi'(z)]e^{2i\theta} \quad (1.29)$$

where θ is the angle between (x, y) and (x', y') coordinate systems. Alternatively, in polar coordinates:

$$\sigma_r + \sigma_\theta = 4\,\text{Re}\{\phi'(z)\}$$
$$\sigma_\theta - \sigma_r + 2i\tau_{r\theta} = 2[\bar{z}\phi''(z) + \psi'(z)]e^{2i\theta}. \quad (1.30)$$

Displacement expressions in terms of ϕ and ψ are given in [4], as:

$$2\mu(u + iv) = \kappa\phi - \overline{z\phi'(z)} - \overline{\psi(z)} \quad (1.31)$$

where

    u is the displacement in the x-direction

    v is the displacement in the y-direction

    $\mu = E/2(1 + v)$

$$\kappa = \begin{cases} (3 - v)/(1 + v), & \text{plane stress} \\ 3 - 4v, & \text{plane strain.} \end{cases}$$

Consider an arc AB with a portion of length ds, n being the outward normal

INTRODUCTORY THEORY OF ELASTICITY

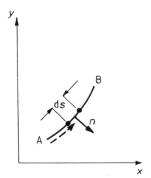

Fig. 1.8 Resultant stress convention.

(Fig. 1.8). If X_n and Y_n are the horizontal and vertical components of the force on ds, then [4]:

$$i \int_s (X_n + iY_n) \, ds = [\phi(z) + z\overline{\phi'(z)} + \overline{\psi(z)}]_A^B \qquad (1.32)$$

Example 1.1

What state of stress is represented by the complex stress functions:

$$\phi(z) = \tfrac{1}{4}T(1+\lambda)z \quad \psi(z) = \tfrac{1}{2}T(1-\lambda)z \ ?$$

Solution: On differentiation and substitution into equations (1.27):

$$\sigma_y = 2\,\mathrm{Re}\{\tfrac{1}{4}T(1+\lambda)\} + \mathrm{Re}\{\tfrac{1}{2}T(1-\lambda)\} = T$$

$$\sigma_x = 2\,\mathrm{Re}\{\tfrac{1}{4}T(1+\lambda)\} - \mathrm{Re}\{\tfrac{1}{2}T(1-\lambda)\} = \lambda T$$

$$\tau_{xy} = \mathrm{Im}\{\tfrac{1}{2}T(1-\lambda)\} = 0$$

which represents a biaxial state of stress applied to an infinite sheet (Fig. 1.9).

Example 1.2

The following complex stress functions are suggested for the solution of a point-loaded wedge problem:

$$\phi(z) = \tfrac{1}{2}C \ln z \quad \psi(z) = -\tfrac{1}{2}C(\ln z + 1). \qquad (1.33)$$

(a) What loading state do the functions represent?
(b) What is the equivalent Airy stress function, F?

14   THE MECHANICS OF FRACTURE AND FATIGUE

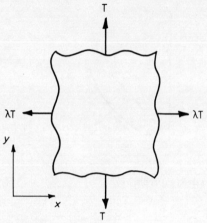

Fig. 1.9 *Biaxial stress state (Example 1.1)*.

Solution: (a) In order to use polar coordinates, re-arrange equations (1.30) to give:

$$\sigma_r = 2\,\text{Re}\{\phi'(z)\} - \text{Re}\{[\bar{z}\phi''(z) + \psi'(z)]e^{2i\theta}\}$$
$$\sigma_\theta = 2\,\text{Re}\{\phi'(z)\} + \text{Re}\{[\bar{z}\phi''(z) + \psi'(z)]e^{2i\theta}\} \quad (1.34)$$
$$\tau_{r\theta} = \text{Im}\{[\bar{z}\phi''(z) + \psi'(z)]e^{2i\theta}\}.$$

Now, noting that:

$$\phi'(z) = \frac{C}{2}\frac{e^{-i\theta}}{r}, \qquad \phi''(z) = -\frac{C}{2}\frac{e^{-2i\theta}}{r^2}$$

$$\psi'(z) = -\frac{C}{2}\frac{e^{-i\theta}}{r}, \qquad \bar{z} = re^{-i\theta}$$

substitute into expressions for σ_r, σ_θ and $\tau_{r\theta}$ to obtain:

$$\sigma_r = \frac{2C}{r}\cos\theta, \quad \sigma_\theta = 0, \quad \tau_{r\theta} = 0.$$

Thus, by considering a wedge of material (Fig. 1.10) the stress functions represent a purely radial stress distribution, the edges of the wedge being free of normal and shear stresses.

In order to calculate the total force acting over an arc, ABC, use is made of the force relationship (equation (1.32)). For the case under consideration this may be expressed as:

$$iX - Y = [\phi(z) + \overline{z\phi'(z)} + \overline{\psi(z)}]_C^A$$

INTRODUCTORY THEORY OF ELASTICITY

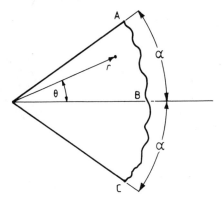

Fig. 1.10 *Wedge geometry.*

where point A has coordinate $re^{i\alpha}$ and point C has coordinate $re^{-i\alpha}$. Relatively straightforward manipulation yields:

$$X = C(2\alpha + \sin 2\alpha), \quad Y = 0.$$

(b) The Airy stress function is given in terms of $\phi(z)$ and $\psi(z)$ by the first of equations (1.26). By substituting from equations (1.33), we obtain:

$$F = \text{Re}\{\bar{z}(\tfrac{1}{2}C \ln z) - \tfrac{1}{2}C(z \ln z - z + z)\} \tag{1.35}$$

where the constant of integration is arbitrarily set to zero, since it will not affect the calculated stress distribution. Recognizing $\ln z = (\ln r + i\theta)$, this reduces to:

$$F = Cr\theta \sin \theta \tag{1.36}$$

which is exactly the function proposed by Boussinesq [3] for the solution of a point-loaded wedge.

1.8 Conformal mapping

By expressing two-dimensional (plane) problems in terms of analytic functions of a complex variable we create the option of applying conformal mapping techniques. These are covered in detail in [4] and [5], however the essential details will be summarized.

Let the z-plane be the physical plane, and R_z a region within that plane. Now introduce an associated plane, the ζ-plane via the transformation:

$$z = \omega(\zeta).$$

Assuming that there is a region in the ζ-plane, R_ζ in which there is a one-for-one correspondence with points in the R_z-plane (Fig. 1.11) the transformation is conformal provided $\omega(\zeta)$ is analytic, and $\omega'(\zeta) \neq 0$ within R_ζ, however singularities

16 THE MECHANICS OF FRACTURE AND FATIGUE

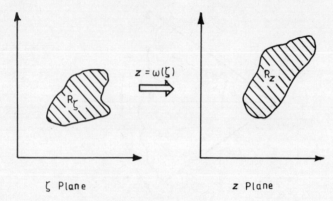

ζ Plane   z Plane

Fig. 1.11 *Conformal mapping between planes.*

in $\omega(\zeta)$ can exist on the boundary of R_ζ in order to model corners. If $\phi(z)$ is analytic in R_z and $\omega(\zeta)$ is a conformal mapping function in R_ζ it may be possible to by-pass determination of $\phi(z)$ and $\psi(z)$ and derive analytic functions $\phi[\omega(\zeta)]$ and $\psi[\omega(\zeta)]$ as functions of ζ. Now define $\phi[\omega(\zeta)]$ as $\phi(\zeta)$ etc., which allows the stress and displacement relationships (equations (1.28) and (1.31)) to be rewritten as:

$$\sigma_x + \sigma_y = 4 \operatorname{Re}\{\phi'(\zeta)/\omega'(\zeta)\}$$
$$\sigma_y - \sigma_x + 2i\tau_{xy} = 2\{\overline{\omega(\zeta)}\,[\phi'(\zeta)/\omega'(\zeta)]' + \psi'(\zeta)\}/\omega'(\zeta) \qquad (1.37)$$
$$2\mu(u + iv) = \kappa\phi(\zeta) - \omega(\zeta)\overline{\phi'(\zeta)/\omega'(\zeta)} - \overline{\psi(\zeta)}$$

When considering the mapping of the region external to the unit circle into the physical region of an elastic material (Fig. 1.12) it can be shown that, provided

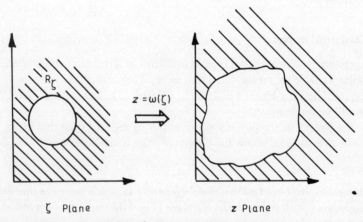

ζ Plane   z Plane

Fig. 1.12 *Conformal mapping of exterior regions.*

INTRODUCTORY THEORY OF ELASTICITY

the finite inner boundary is free of tractions, which implies certain continuity constraints across the boundary, then in the region of the ζ (parameter) plane external to the unit circle [4, 6]:

$$\psi(\zeta) = -\bar{\phi}(1/\zeta) - \bar{\omega}(1/\zeta) \frac{\phi'(\zeta)}{\omega'(\zeta)}. \tag{1.38}$$

(Readers may note the similarity in form between this relationship and equation (1.32) in the special case in which the contour is traction free.) Thus, somewhat surprisingly, the problem is reduced to the determination of one analytic function, since $\psi(\zeta)$ is now expressible in terms of $\phi(\zeta)$. The mapping concepts as applied to problems in elasticity, are somewhat overawing at first sight. In order to clarify each step in the process a straightforward problem is tackled by the conformal mapping technique.

1.8.1 *Elliptical cutout in an infinite sheet under uniaxial tension*

The transformation

$$z = \omega(\zeta) = c\left(\zeta + \frac{m}{\zeta}\right) \quad m < 1, \tag{1.39}$$

where $c = (a + b)/2$ and $m = (a - b)/(a + b)$, maps to the region external to an ellipse of major axis $2a$, minor axis $2b$ in the z (physical) plane from the region outside a unit circle in the ζ-plane (Fig. 1.13). The transformation is not conformal at points where $\omega'(\zeta) = 0$, i.e. at $\zeta = \pm m^{1/2}$. However, these points are inside the unit circle, and hence are not contained in the region from which the physical region is mapped. Results obtained will not be invalidated by lack of conformality. Note also that $x = a \cos \theta$, $y = b \sin \theta$ where θ is shown in Fig. 1.13.

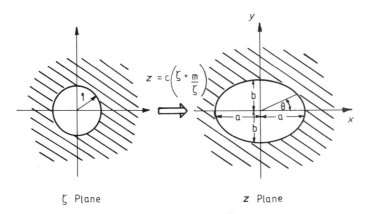

Fig. 1.13 *Conformal mapping of unit circle to ellipse.*

In the z-plane, the conditions for large $|z|$ must conform with those obtained in Example 1.1 for the case $\lambda = 0$. Thus:

$$\lim_{|z| \to \infty} \phi(z) = \tfrac{1}{4}Tz, \qquad \lim_{|z| \to \infty} \psi(z) = \tfrac{1}{2}Tz. \tag{1.40}$$

Similarly, for large $|\zeta|$, since z is defined by equation (1.39):

$$\lim_{|\zeta| \to \infty} \phi(\zeta) = \tfrac{1}{4}TC\zeta, \qquad \lim_{|\zeta| \to \infty} \psi(\zeta) = \tfrac{1}{2}TC\zeta \tag{1.41}$$

The absence of tractions on the unit circle is ensured by the form of equation (1.38). Taking a general form of $\phi(\zeta)$ for the whole of R_ζ as:

$$\phi(\zeta) = \tfrac{1}{4}TC\zeta + \sum_{1}^{\infty} A_n \zeta^{1-2n} \tag{1.42}$$

where coefficients A_n are real. Even powers of ζ are eliminated on the basis of the biaxial symmetry of the problem, other positive powers of ζ are excluded because of the requirement to produce only a resultant $\tfrac{1}{4}TC\zeta$ at large $|\zeta|$.

It is now a straightforward procedure to differentiate and substitute from equations (1.39) and (1.42) into (1.38). After some manipulation this yields:

$$\psi(\zeta) = -TC/4\zeta - \sum_{1}^{\infty} A_n \zeta^{2n-1}$$

$$-[(m\zeta^2 + 1)/(\zeta^2 - m)] \left(TC\zeta/4 + \sum_{1}^{\infty} A_n(1 - 2n)\zeta^{1-2n} \right). \tag{1.43}$$

Now, remembering the condition that $\psi(\zeta) \to \tfrac{1}{2}TC\zeta$ for large $|\zeta|$ the above expression becomes:

$$TC\zeta/2 = -\sum_{1}^{\infty} A_n \zeta^{2n-1} - m\left(TC\zeta/4 + \sum_{1}^{\infty} A_n(1 - 2n)\zeta^{1-2n} \right). \tag{1.44}$$

Recalling the restrictions on positive powers of ζ, the only possible value is $n = 1$. This yields:

$$TC\zeta/2 = -A_1 \zeta - (mTC\zeta/4) + A_1 \zeta^{-1}$$

or

$$A_1 = -\tfrac{1}{4}TC(2 + m).$$

Thus

$$\phi(\zeta) = \tfrac{1}{4}TC[\zeta - (2 + m)\zeta^{-1}]. \tag{1.45}$$

It is instructive, but somewhat time consuming, to substitute this result into

INTRODUCTORY THEORY OF ELASTICITY

equation (1.38) to obtain:

$$\psi(\zeta) = \frac{TC}{2}\left(\frac{\zeta^3 - \zeta(m^2 + 2m + 1) - \zeta^{-1}}{\zeta^2 - m}\right). \tag{1.46}$$

In deriving the stresses around the ellipse, consider a small element of material at the boundary (Fig. 1.14) where the stresses acting on this element are designated

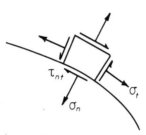

Fig. 1.14 *Stress element on ellipse boundary.*

σ_n, σ_t and τ_{nt}. The value of $(\sigma_n + \sigma_t)$ is obtained from the first of equations (1.37), recognizing that the transformation does not affect the sum of the direct stresses, thus:

$$(\sigma_n + \sigma_t) = 4\,\text{Re}\{\phi'(\zeta)/\omega'(\zeta)\}. \tag{1.47}$$

However, for the case of a traction-free boundary, $\sigma_n = \tau_{nt} = 0$, which gives:

$$\sigma_t = 4\,\text{Re}\{\phi'(\zeta)/\omega'(\zeta)\} \tag{1.48}$$

by substitution from equations (1.39) and (1.46) this becomes:

$$\frac{\sigma_t}{T} = \frac{1 - 2m - m^2 + 2\cos 2\theta}{1 + m^2 - 2m\cos 2\theta} \tag{1.49}$$

which represents the stress concentration factor at points on the elliptic hole. Values of σ_t/T are shown in Fig. 1.15 in the range $0 \leq \theta \leq \pi/2$ for various values of a/b.

Attention is now focused on the point on the hole boundary at $\theta = 0$. Substituting $\theta = 0$, $m = (a - b)/(a + b)$ into equation (1.49) produces:

$$\sigma_t/T = 1 + 2a/b, \tag{1.50}$$

which indicates that the degeneration of the ellipse into a crack as $b \to 0$ produces an infinite stress concentration at the crack tip. An alternative form of equation (1.50) is obtained by substituting the known expression for root radius of an ellipse ρ, namely:

$$\rho = b^2/a$$

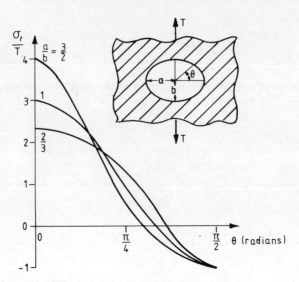

Fig. 1.15 *Stress concentration around elliptical cutout.*

in order to yield:

$$\frac{\sigma_t}{T} = 1 + 2\left(\frac{a}{\rho}\right)^{1/2} \tag{1.51}$$

References

1. Timoshenko S. P. and Goodier J. N. (1970), *Theory of Elasticity*, 3rd Edn, McGraw-Hill, New York.
2. Dugdale D. S. (1968), *Elements of Elasticity*, Pergamon, Oxford.
3. Boussinesq J. (1885), *Application des Potentiels a l'Etude ... Elastiques*, Gauthier-Villars, Paris.
4. Muskhelishvili N. I. (1953), *Some Basic Problems of the Mathematical Theory of Elasticity*, Noordhoff, Leiden
5. Godfrey D. E. R. (1959), *Theoretical Elasticity and Plasticity for Engineers*, Thames and Hudson, London.
6. Bowie O. L. (1973), 'Solutions of plane crack problems by mapping technique in *Methods of Analysis and Solutions of Crack Problems*, Ed. G. C. Sih, Noordhoff, Leiden.

2 Energy Considerations

2.1 Introduction

In Chapter 1 we noted that stresses at the end of the major axis of an elliptical cutout within a loaded body tend to infinity as the ellipse degenerates into a crack. In order to understand why such cracked components are capable of withstanding loads up to a certain, critical level, it is necessary to study the energy changes during crack extension.

Additionally, the change in stiffness which occurs as a result of crack extension is associated with the energetics of the system, and will be a useful concept in the understanding of both experimental and analytical methods.

2.2 General

When a body is loaded, the movement of the applied loads does work on the body, which is stored in the form of strain energy. It is possible to express the energy stored per unit volume of material, U, in terms of stress components alone [1]. Then, for a general three-dimensional loading system:

$$U = \frac{1}{2E} [\sigma_x^2 + \sigma_y^2 + \sigma_z^2 - 2v(\sigma_x\sigma_y + \sigma_y\sigma_z + \sigma_z\sigma_x)] \\ + 2(1+v)(\tau_{xy}^2 + \tau_{yz}^2 + \tau_{zx}^2). \tag{2.1}$$

In the case of simple uniaxial tension, σ, the above expression reduces to:

$$U = \sigma^2/2E. \tag{2.2}$$

2.3 The Griffith crack

An understanding of the stress field equations for (say) an elliptical cutout indicates that sharp notches are capable of producing very large stress concentrations. Nevertheless, infinitely sharp flaws (cracks) may be present within a loaded body without producing failure of the component. In order to understand this apparent paradox, Griffith [2] confined his attention to a brittle material containing a single crack of length $2a$ (Fig. 2.1) and considered the energy changes in the system associated with an incremental extension of the

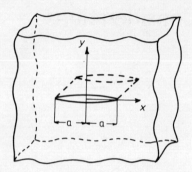

Fig. 2.1 *Griffith crack.*

crack. However, it is necessary to emphasize that two conditions must be fulfilled for extension of a notch:

(a) The stresses ahead of the notch must reach a critical value.
(b) The total energy of the system must be reduced by an incremental extension of the notch.

2.3.1 *Potential energy*

At this point we simplify Griffith's approach (although his complete solution will be derived in Chapter 3). Firstly, consider the plate in the absence of the crack when the material is uniformly stressed and clamped remotely (Fig. 2.2(a)). The energy per unit volume is known to be $\sigma^2/2E$. If we now introduce a crack

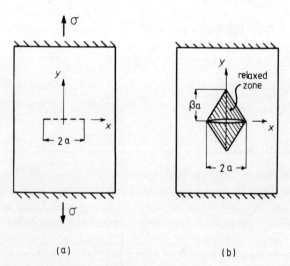

Fig. 2.2 *Approximate energy model.*

ENERGY CONSIDERATIONS

of length $2a$ into the plate there will be a general relaxation of the material above and below the crack, and some strain energy will be released. In order to obtain an approximate solution, let us assume that the relaxed zone is in the form of a triangle, height βa (Fig. 2.2(b)). The relaxed volume of the shaded zone to the right of the y-axis is given by:

$$\beta a^2 t$$

where t is the plate thickness. Hence, the energy, U, released per unit thickness is given by:

U = energy per unit volume × volume

$$= \frac{\sigma^2}{2E} \beta a^2 \qquad (2.3)$$

this is in good accord with Griffith's accurate solution for plane stress:

$$U = \frac{\sigma^2}{2E} \pi a^2, \quad \frac{\partial U}{\partial a} = \frac{\sigma^2}{E} \pi a \qquad (2.4)$$

whilst, for plane strain (since $\epsilon = (1 - v^2)\sigma/E$),

$$U = \frac{\sigma^2}{2E} \pi a^2 (1 - v^2), \quad \frac{\partial U}{\partial a} = \frac{\sigma^2 \pi a}{E}(1 - v^2) \qquad (2.5)$$

The variation of U with crack length is shown schematically in Fig. 2.3. Since U represents a release of energy, it is plotted as a negative quantity.

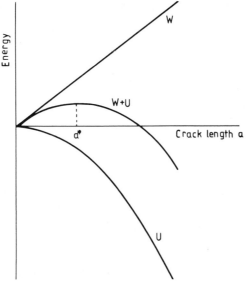

Fig. 2.3 *Variation of energy with crack length.*

2.3.2 Surface energy

Now, whilst energy is released as a result of the relaxation process, an energy input, W, is required to produce the crack growth at each tip. For our purposes it is sufficient to assume that the energy required for each equal increment of crack length is a constant. This is equivalent to saying that the energy required to break one atomic bond ahead of the crack is the same as that required to break the next, and so on. Hence W increases linearly with increasing crack length (Fig. 2.3). W takes a positive value since it represents an energy input to the system, which is employed in the creation of new crack surfaces.

2.3.3 Total energy

By summing the separate energies at each crack length we obtain the curve representing total energy of the system (Fig. 2.3). It is evident that increasing crack length in the region $0 \leqslant a \leqslant a^*$ requires an energy input to the system. However, for $a > a^*$ energy is released as a result of crack extension, and the crack will propagate.

2.3.4 Energy release rates

Clearly crack instability is associated with the stationary value of the total energy curve, beyond this point the energy released during an incremental crack extension exceeds the energy required to create new crack surfaces. It is the strain energy release rate on which we now focus our attention.

The value of $(\partial U/\partial a)$ defines the strain energy release rate for an incremental crack extension. The elastic energy release rate per tip (G) is defined by:

$$G = \partial U/\partial a$$

thus, from equations (2.4) and (2.5):

$$G = \begin{cases} \dfrac{\pi}{E} \sigma^2 a, & \text{plane stress} \qquad (2.6) \\[1em] \dfrac{\pi}{E} \sigma^2 a (1 - v^2), & \text{plane strain.} \qquad (2.7) \end{cases}$$

The value $(\partial W/\partial a)$ defines the energy absorbed during an incremental crack extension, and is normally assigned the symbol R. Thus the threshold condition for unstable crack growth may be expressed as:

$$\partial U/\partial a = \partial W/\partial a \quad \text{or} \quad G = R.$$

However, since R is a constant, we may define a constant, critical value of G, namely G_{crit}, at which unstable crack propagation will occur. At this stage we

ENERGY CONSIDERATIONS

anticipate that G_{crit} will be a material property, since it depends on the energy required to break atomic bonds within the material.

It is possible to determine G_{crit} via equations (2.6) and (2.7). By testing a specimen containing a crack of length $2a$, and noting the stress level at failure, σ_c, the value of G_{crit} may be derived as:

$$G_{crit} = \begin{cases} G_c = \dfrac{\sigma_c^2 \pi a}{E}, & \text{plane stress} \quad (2.8) \\[2mm] G_{IC} = \dfrac{\sigma_c^2 \pi a}{E}(1 - v^2), & \text{plane strain.} \quad (2.9) \end{cases}$$

Re-arrangement of these equations yields:

$$(G_c E)^{1/2} = \sigma_c (\pi a)^{1/2}, \qquad \text{plane stress} \quad (2.10)$$

$$(G_{IC} E)^{1/2} = \sigma_c (\pi a)^{1/2} (1 - v^2)^{1/2}, \qquad \text{plane stress.} \quad (2.11)$$

Note that the above equations contain only material properties on the left-hand side, and geometrical and loading parameters on the right-hand side.

2.4 Compliance

A load P is applied to a cracked body (Fig. 2.4). The associated displacement at the loading point in the direction of application of the load is related to the load by:

force = stiffness × displacement

or

$$P = (1/C)u \qquad (2.12)$$

the term C, being the reciprocal of stiffness, is termed the compliance. Rewriting

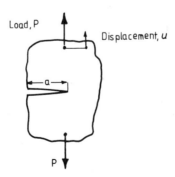

Fig. 2.4 *Cracked body.*

the equation in the usual form:

$$u = CP. \tag{2.13}$$

In general, C will be a function of the geometry of the body, including the crack length. However, when considering infinitesimal changes in crack length, C is considered to be the same for crack lengths a and $(a + \delta a)$. Thus we rewrite equation (2.13) as:

$$\delta u = C \, \delta P \tag{2.14}$$

2.4.1 Loading conditions

In Section 2.3 only so-called fixed-grip conditions were considered (see Fig. 2.5(a)). This implies that energy for crack extension must be supplied by released elastic energy alone. In this case there will be a small reduction in the remotely applied load, δP associated with an increment in crack length δa, and a maintained displacement u. The strain energy release with fixed grips is $-\tfrac{1}{2} u \, \delta P$. Apply equation (2.13) to the above relationship to obtain:

$$-\tfrac{1}{2} u \, \delta P = -\tfrac{1}{2} CP \, \delta P. \tag{2.15}$$

Consider now the case of fixed-load conditions (Fig. 2.5(b)). If the loading points are free to move during crack extension, work will be done by these loads. In this case the load is maintained during an increase in crack length, and a displacement, δu, is produced. The energy release associated with these fixed-load conditions is given by:

— work done + increase in strain energy of body

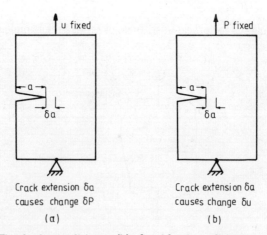

Fig. 2.5 (a) *Fixed-grip conditions;* (b) *fixed-load conditions.*

ENERGY CONSIDERATIONS

or

$$-P\,\delta u + \tfrac{1}{2}P\,\delta u = -\tfrac{1}{2}P\,\delta u \tag{2.16}$$

employing equation (2.14) this becomes:

$$-\tfrac{1}{2}P\,\delta u = -\tfrac{1}{2}CP\,\delta P. \tag{2.17}$$

It is possible to argue the above relationships graphically (see [3] and [4]).

By comparison of equations (2.15) and (2.17) it is noted that, for a small crack extension, the reduction in stored elastic energy under fixed-grip conditions is the same as the reduction in energy under constant-load conditions.

Compliance concepts will be mentioned again with reference to numerical and experimental methods in fracture mechanics, and the derivation of weight functions in Chapter 4.

References

1. Dugdale D. S. (1968), *Elements of Elasticity*, Pergamon, Oxford.
2. Griffith A. A. (1921), 'The phenomena of rupture and flow in solids', *Phil. Trans. R. Soc., Lond.* A, **221**, 163–97.
3. Broek D. (1974), *Elementary Engineering Fracture Mechanics*, Noordhoff, Leiden.
4. Knott J. F. (1973), *Fundamentals of Fracture Mechanics*, Butterworth, Sevenoaks.

3 Stresses and Displacements in Cracked Bodies

3.1 Introduction

In the previous chapters we pursued the theory of elasticity to the point at which it was evident that stresses at the tip of an elliptical cutout tend to infinity as the ellipse becomes infinitely sharp. Furthermore, we have seen that energy is released during crack extension, and that a critical value of this energy release rate can be obtained directly from geometrical and loading parameters via equations (2.6).

We now consider the stress and displacement field equations for various configurations in the vicinity of the crack tip in order to demonstrate that the energy release rate may be derived directly from the magnitude of the stress field singularity, termed the stress intensity factor. The attainment of a critical energy release rate may be associated with a critical value of stress intensity factor, the fracture toughness.

3.2 Stresses in cracked bodies

Consider the complex stress functions derived for an elliptical cutout in an infinite sheet (equations (1.45) and (1.46)). With remote loading σ, allowing the minor axis, b, to tend to zero we obtain:

$$\phi(\zeta) = \frac{\sigma a}{8}\left(\zeta - \frac{3}{\zeta}\right) \tag{3.1}$$

$$\psi(\zeta) = \frac{\sigma a}{4}\left(\frac{\zeta^3 - 4\zeta - \zeta^{-1}}{\zeta^2 - 1}\right) \tag{3.2}$$

in order to convert back to z-plane coordinates, note that:

$$\phi(\zeta) = -\frac{\sigma a}{8}\left(\zeta + \frac{1}{\zeta}\right) + \frac{\sigma a}{4}\left(\zeta - \frac{1}{\zeta}\right) \tag{3.3}$$

but

$$z = \frac{a}{2}\left(\zeta + \frac{1}{\zeta}\right), \quad (z^2 - a^2)^{1/2} = \frac{a}{2}\left(\zeta - \frac{1}{\zeta}\right). \tag{3.4}$$

Thus

$$\phi(z) = \tfrac{1}{4}\sigma[2(z^2 - a^2)^{1/2} - z]. \qquad (3.5)$$

Following a similar approach, equation (3.2) may be rewritten as:

$$\phi(z) = \tfrac{1}{2}\sigma[z - a^2(z^2 - a^2)^{-1/2}]. \qquad (3.6)$$

$\phi(z)$ and $\psi(z)$ are the stress functions for a remotely loaded crack in an infinite sheet (Fig. 3.1).

Note: It is a valuable manipulative exercise to confirm, via equations (1.27), that the stress functions of equations (3.5) and (3.6) do in fact satisfy the boundary conditions for a stress-free crack in a remotely loaded sheet, namely:

$$\lim_{|z| \to \infty} \sigma_y = \sigma, \quad \lim_{|z| \to \infty} \sigma_x = 0, \quad \lim_{|z| \to \infty} \tau_{xy} = 0$$

whilst:

$$\sigma_y = \tau_{xy} = 0, \quad -a \leqslant x \leqslant a.$$

By a shift of the (x, y) axes to (x', y), centred on the right-hand crack tip (Fig. 3.1) the expression for σ_y, namely:

$$\sigma_y = \sigma x(x^2 - a^2)^{-1/2}, \quad |x| \geqslant a \qquad (3.7)$$

becomes:

$$\sigma_y = \sigma(a + x')(x'^2 + 2ax')^{-1/2}, \quad |x| \geqslant a \qquad (3.8)$$

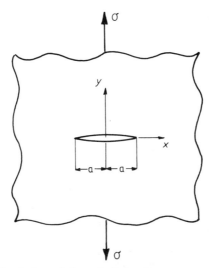

Fig. 3.1 *Remotely loaded crack in an infinite sheet.*

and

$$\lim_{x' \to 0} \sigma_y = \sigma\left(\frac{a}{2x'}\right)^{1/2}. \tag{3.9}$$

Evidently, as the tip is approached, $\sigma_y \to \infty$ as $1/(\text{distance from tip})^{1/2}$.

3.3 Stress intensity factor

In general, Sneddon [1] has shown that, in the vicinity of the crack tip:

$$\sigma_x = \frac{K_\mathrm{I}}{(2\pi r)^{1/2}} \cos\left(\frac{\theta}{2}\right)\left[1 - \sin\left(\frac{\theta}{2}\right)\sin\left(\frac{3\theta}{2}\right)\right] + \text{non-singular terms}$$

$$\sigma_y = \frac{K_\mathrm{I}}{(2\pi r)^{1/2}} \cos\left(\frac{\theta}{2}\right)\left[1 + \sin\left(\frac{\theta}{2}\right)\sin\left(\frac{3\theta}{2}\right)\right] + \text{non-singular terms} \tag{3.10}$$

$$\tau_{xy} = \frac{K_\mathrm{I}}{(2\pi r)^{1/2}} \cos\left(\frac{\theta}{2}\right) \sin\left(\frac{\theta}{2}\right) \cos\left(\frac{3\theta}{2}\right) + \text{non-singular terms}$$

where (r, θ) are polar coordinates based on the right-hand crack tip, and σ_x, σ_y, τ_{xy} are the stresses acting on the element so located (Fig. 3.2). K_I is given by:

$$K_\mathrm{I} = \sigma(\pi a)^{1/2}. \tag{3.11}$$

K_I defines the *magnitude* of the crack tip stress field singularity, and is termed the stress intensity factor [2]. In fact we note that there are other, equivalent ways of defining K_I, namely:

$$K_\mathrm{I} = \lim_{z \to a} (2\pi)^{1/2}(z - a)^{1/2}\sigma_y \tag{3.12}$$

and

$$K_\mathrm{I} = \lim_{z \to a} 2(2\pi)^{1/2}(z - a)^{1/2}\phi'(z). \tag{3.13}$$

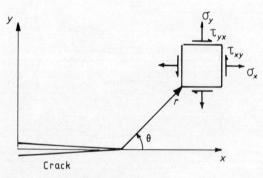

Fig. 3.2 *Crack tip coordinate and stress system.*

STRESSES AND DISPLACEMENTS IN CRACKED BODIES

Recalling equations (2.6) we see:

$$K_\mathrm{I} = \begin{cases} (GE)^{1/2}, & \text{plane stress} \quad (3.14) \\ \left(\dfrac{GE}{1-v^2}\right)^{1/2}, & \text{plane strain.} \quad (3.15) \end{cases}$$

Thus K_I is a function of the energy release rate during an incremental crack extension.

3.4 Fracture toughness

Unstable crack propagation will occur when K_I reaches a critical value K_C, termed the fracture toughness. The value of K_C depends on the amount of crack tip constraint, and is therefore a function of specimen thickness and geometry. For the symmetrical configuration considered, under maximum crack tip constraint (plane strain conditions), the critical stress intensity factor is designated K_IC, the plane strain fracture toughness. This is the minimum value of fracture toughness, and is considered in more detail in Chapter 6.

Fracture toughness values may be obtained via well-defined test procedures (see Chapter 6). At this stage it is sufficient for the reader to note that there will be some specific, critical value of stress intensity for a given material and configuration.

Example 3.1

The stress functions for a crack in an infinite sheet, subjected to remote uniaxial tension are:

$$\phi_a(z) = \frac{\sigma}{4}[2(z^2 - a^2)^{1/2} - z] \tag{3.16}$$

$$\psi_a(z) = \frac{\sigma}{2}\left(z - \frac{a^2}{(z^2 - a^2)^{1/2}}\right). \tag{3.17}$$

(a) By superposition, derive stress functions for a pressurized crack in an infinite sheet.
(b) Evaluate stresses from these new stress functions to prove that the superposition is correct.
(c) Determine the stress intensity factor for the pressurized crack.

Solution: (a) The necessary superposition is illustrated in Fig. 3.3. The stress functions associated with configuration (a) are given in the question. Those associated with configuration (b) are obtained from equations (3.16) and (3.17) by substituting $a = 0$, $\sigma = -\sigma$, giving:

$$\phi_b(z) = -\sigma z/4, \quad \psi_b(z) = -\sigma z/2. \tag{3.18}$$

32   THE MECHANICS OF FRACTURE AND FATIGUE

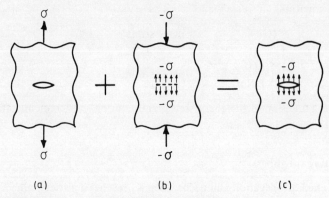

Fig. 3.3 *Superposition technique (Example 3.1)*.

Addition of ϕ_a and ϕ_b, and of ψ_a and ψ_b yields:

$$\phi_c(z) = \tfrac{1}{2}\sigma[(z^2 - a^2)^{1/2} - z] \tag{3.19}$$

$$\psi_c(z) = -\tfrac{1}{2}\sigma[a^2(z^2 - a^2)^{-1/2}] \tag{3.20}$$

(b) In order to derive stresses, evaluate derivatives:

$$\phi'_c(z) = \tfrac{1}{2}\sigma[z(z^2 - a^2)^{-1/2} - 1] \tag{3.21}$$

$$\psi'_c(z) = \tfrac{1}{2}\sigma[a^2 z(z^2 - a^2)^{-3/2}] \tag{3.22}$$

$$\phi''_c(z) = \tfrac{1}{2}\sigma[-a^2(z^2 - a^2)^{-3/2}]. \tag{3.23}$$

Now, from equation (1.27) we find that for large $|z|$:

$$\sigma_y = 0, \quad \sigma_x = 0, \quad \tau_{xy} = 0 \tag{3.24}$$

and that along the x-axis:

$$\sigma_y = -\sigma = p \text{ (say)}, \quad -a \leqslant x \leqslant a$$

$$\tau_{xy} = 0, \quad -\infty \leqslant x \leqslant +\infty$$

where p represents a constant pressure. Thus the stress functions define the state of stress in an infinite sheet containing a single pressurized crack.

(c) Substituting from equation (3.21) into (3.13) we obtain an expression for the stress intensity factor, namely:

$$K_I = \sigma(\pi a)^{1/2}$$

Note two points of some importance:

   (a) The stress intensity factor for a crack in a sheet under remote uniaxial

stress is the same as the stress intensity factor for the crack subjected to a pressure of the same magnitude.
(b) In deriving stress intensity factors we have the choice of *either* solving the problem of the stress-free crack under remote loading *or* of a loaded crack without other boundary tractions, where the crack loading is equal and opposite to the stresses which exist in the equivalent remotely loaded, uncracked configuration.

3.5 Crack shape

Recalling the expression for displacements in terms of stress functions (equation (1.31)):

$$2\mu(u + iv) = \kappa\phi(z) - z\overline{\phi'(z)} - \overline{\psi(z)} \qquad (3.25)$$

whilst for the remotely loaded crack from equations (3.5) and (3.6):

$$\phi'(z) = \tfrac{1}{2}T[(z^2 - a^2)^{-1/2}z - \tfrac{1}{2}]$$
$$\overline{\phi'(z)} = \tfrac{1}{2}T[(z^2 - a^2)^{-1/2}z - \tfrac{1}{2}] \qquad (3.26)$$
$$\overline{\psi(z)} = \tfrac{1}{2}T[z - a^2(z^2 - a^2)^{-1/2}].$$

Now, along the x-axis:

$$\overline{(z^2 - a^2)^{-1/2}} = (x^2 - a^2)^{-1/2} = i(a^2 - x^2)^{-1/2}, \quad -a \leqslant x \leqslant a. \qquad (3.27)$$

Hence, substituting from equations (3.5) and (3.26) into (3.25) and taking the imaginary part of the resulting expression gives:

$$2\mu v = \tfrac{1}{2}\sigma(\kappa + 1)(a^2 - x^2)^{1/2} \qquad (3.28)$$

$$v = \sigma\left(\frac{\kappa + 1}{4\mu}\right)(a^2 - x^2)^{1/2} \qquad (3.29)$$

which indicates that the crack opens to form an ellipse, semi-major axis, a, semi-minor axis, $\sigma a(\kappa + 1)/4\mu$.

In order to examine the limiting values of crack opening we again shift the (x, y) axes to (x', y) centred on the right-hand crack tip (Fig. 3.1). Replacing x by $(a + x')$ in equation (3.29):

$$v = \sigma\left(\frac{\kappa + 1}{4\mu}\right)(-x'^2 - 2ax')^{1/2} \qquad (3.30)$$

and as x' becomes very small, we obtain:

$$v = \sigma\left(\frac{\kappa + 1}{4\mu}\right)(-2ax')^{1/2} \qquad (3.31)$$

or

$$v = K_I \left(\frac{\kappa + 1}{2\mu}\right)\left(-\frac{x'}{2\pi}\right)^{1/2} \quad (3.32)$$

where K_I is defined by equation (3.11).

Thus, as the crack tip is approached from the negative x'-direction, $v \to 0$ as (distance from tip)$^{1/2}$. In general, the form of the displacement field is given by [3]:

$$u = \left(\frac{K_I}{4\mu}\right)\left(\frac{r}{2\pi}\right)^{1/2}\left[(2\kappa - 1)\cos\left(\frac{\theta}{2}\right) - \cos\left(\frac{3\theta}{2}\right)\right] + \cdots$$

$$v = \left(\frac{K_I}{4\mu}\right)\left(\frac{r}{2\pi}\right)^{1/2}\left[(2\kappa + 1)\sin\left(\frac{\theta}{2}\right) - \sin\left(\frac{3\theta}{2}\right)\right] + \cdots \quad (3.33)$$

3.6 Energy released

Equation (2.2) gives the strain energy released per unit volume in terms of stress and strain. In the case of a single crack in an infinite sheet, subjected to remote uniaxial tension σ, stress-free surfaces have been produced by introducing stresses which cancel out those along the crack line in the unflawed sheet (Fig. 3.4).

The difference between the strain energy of the uncracked and the cracked sheets is simply the strain energy of Fig. 3.4(b). Assuming a sheet of unit thickness, this is given by:

$$U = \tfrac{1}{2}\int_a p(x)\,v(x, a)\,dx \quad (3.34)$$

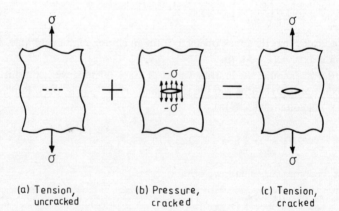

(a) Tension, uncracked   (b) Pressure, cracked   (c) Tension, cracked

Fig. 3.4 *Superposition for energy release calculation.*

STRESSES AND DISPLACEMENTS IN CRACKED BODIES 35

where $p(x)$ is the loading distribution along the crack line, and $v(x, a)$ the vertical opening of a crack of length $2a$, at x. $\int_a$ represents the integral over the whole of the crack surface.

Substituting for $v(x, a)$ from equation (3.29), together with $p(x) = -\sigma$, we obtain:

$$U = \tfrac{1}{2}\sigma^2 \left(\frac{\kappa + 1}{4\mu}\right) \cdot 2\int_{-a}^{a} (a^2 - x^2)^{1/2} \, dx \tag{3.35}$$

where the factor 2 before the integral accounts for both upper and lower crack surfaces. Finally, the integral gives:

$$U = \sigma^2 \left(\frac{\kappa + 1}{8\mu}\right) \pi a^2 \tag{3.36}$$

which conforms with Griffith's solution, equation (2.4) for plane stress, and equation (2.5) for plane strain.

3.7 Modes of crack tip deformation

Thus far we have assumed that loading and geometry are symmetrical about the crack line. This mode of deformation is a common one; however, a crack may be deformed in any of three independent ways (Fig. 3.5) namely:

(a) Opening mode (mode I), having symmetry about (x, y) and (x, z) planes.

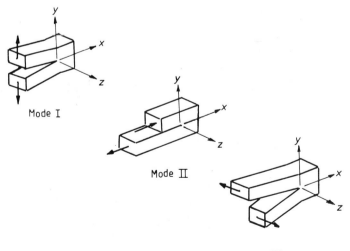

Fig. 3.5 *Modes of crack deformation.*

(b) Sliding, or shear mode (mode II), having anti-symmetry about the (x, z) plane, and symmetry about the (x, y) plane.
(c) Tearing mode (mode III), having anti-symmetry about (x, y) and (x, z) planes.

(*Note*: In this section the z suffix refers to the Cartesian coordinate system, and *not* the complex number $(x + iy)$).

In a particular problem the crack tip stress and displacement fields may be treated as any one, or by linear superposition, any combination of the individual modes. Via suitable stress functions [4], we may develop expressions for each of the three modes of deformation. Clearly, there will be a stress intensity factor $K_N (N = I, II, III)$, defining the magnitude of the stress field singularity, associated with each of the three modes of crack tip deformation. The full set of stress and displacement fields are:

Mode I (plane strain)

$$\sigma_x = \frac{K_I}{(2\pi r)^{1/2}} \cos\left(\frac{\theta}{2}\right) \left[1 - \sin\left(\frac{\theta}{2}\right) \sin\left(\frac{3\theta}{2}\right)\right]$$

$$\sigma_y = \frac{K_I}{(2\pi r)^{1/2}} \cos\left(\frac{\theta}{2}\right) \left[1 + \sin\left(\frac{\theta}{2}\right) \sin\left(\frac{3\theta}{2}\right)\right]$$

$$\tau_{xy} = \frac{K_I}{(2\pi r)^{1/2}} \sin\left(\frac{\theta}{2}\right) \cos\left(\frac{\theta}{2}\right) \cos\left(\frac{3\theta}{2}\right) \quad (3.37)$$

$$\sigma_z = v(\sigma_x + \sigma_y), \quad \tau_{xz} = \tau_{yz} = 0$$

$$u = \frac{K_I}{G} \left(\frac{r}{2\pi}\right)^{1/2} \cos\left(\frac{\theta}{2}\right) \left[1 - 2v + \sin^2\left(\frac{\theta}{2}\right)\right]$$

$$v = \frac{K_I}{G} \left(\frac{r}{2\pi}\right)^{1/2} \sin\left(\frac{\theta}{2}\right) \left[2 - 2v - \cos^2\left(\frac{\theta}{2}\right)\right]$$

$$w = 0$$

where G is the shear modulus of elasticity and

$$K_I = (2\pi r)^{1/2} \lim_{r \to 0} \sigma_y \quad (3.38)$$

Mode II (plane strain):

$$\sigma_x = -\frac{K_{II}}{(2\pi r)^{1/2}} \sin\left(\frac{\theta}{2}\right) \left[2 + \cos\left(\frac{\theta}{2}\right) \cos\left(\frac{3\theta}{2}\right)\right]$$

$$\sigma_y = \frac{K_{II}}{(2\pi r)^{1/2}} \sin\left(\frac{\theta}{2}\right) \cos\left(\frac{\theta}{2}\right) \cos\left(\frac{3\theta}{2}\right)$$

$$\tau_{xy} = \frac{K_{II}}{(2\pi r)^{1/2}} \cos\left(\frac{\theta}{2}\right) \left[1 - \sin\left(\frac{\theta}{2}\right) \sin\left(\frac{3\theta}{2}\right)\right] \quad (3.39)$$

$$\sigma_z = v(\sigma_x + \sigma_y), \quad \tau_{xz} = \tau_{yz} = 0$$

$$u = \frac{K_{II}}{G} \left(\frac{r}{2\pi}\right)^{1/2} \sin\left(\frac{\theta}{2}\right) \left[2 - 2v + \cos^2\left(\frac{\theta}{2}\right)\right]$$

$$v = \frac{K_{II}}{G} \left(\frac{r}{2\pi}\right)^{1/2} \cos\left(\frac{\theta}{2}\right) \left[-1 + 2v + \sin^2\left(\frac{\theta}{2}\right)\right]$$

$$w = 0$$

where

$$K_{II} = (2\pi r)^{1/2} \lim_{r \to 0} \tau_{xy}. \quad (3.40)$$

To convert the above expressions for plane stress, set $\sigma_z = 0$, and replace v by $v/(1 + v)$.

Mode III:

$$\tau_{xz} = -\frac{K_{III}}{(2\pi r)^{1/2}} \sin\left(\frac{\theta}{2}\right)$$

$$\tau_{yz} = \frac{K_{III}}{(2\pi r)^{1/2}} \cos\left(\frac{\theta}{2}\right)$$

$$\sigma_x = \sigma_y = \sigma_z = \tau_{xy} = 0 \quad (3.41)$$

$$w = \frac{K_{III}}{G} \left(\frac{2r}{\pi}\right)^{1/2} \sin\left(\frac{\theta}{2}\right)$$

$$u = v = 0$$

where

$$K_{III} = (2\pi r)^{1/2} \lim_{r \to 0} \tau_{zy}. \quad (3.42)$$

The stress components and (r, θ) coordinate system are shown in Fig. 3.6. u, v and w are the displacements in the x-, y- and z-directions respectively. Equations

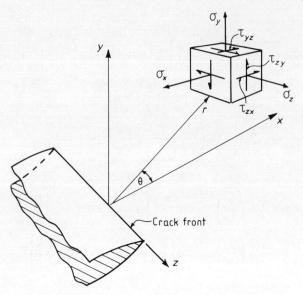

Fig. 3.6 *Three-dimensional crack tip coordinate and stress systems.*

(3.37), (3.39) and (3.41) are exact only in the limit as r approaches zero, but will represent a good approximation in the region where $r \ll a$ and $r \ll L$, where L represents the smallest planar dimension of the body.

The strain energy release rate (energy per unit crack extension) associated with an incremental crack extension was derived in Section 3.3 for the case of mode I loading:

$$G_\text{I} = \begin{cases} K_\text{I}^2/E, & \text{plane stress} \\ (1-v^2)K_\text{I}^2/E, & \text{plane strain.} \end{cases}$$ (3.43)
(3.44)

There will be analogous expressions for energy release under mode II crack extension:

$$G_\text{II} = \begin{cases} K_\text{II}^2/E, & \text{plane stress} \\ (1-v^2)K_\text{II}^2/E, & \text{plane strain} \end{cases}$$ (3.45)
(3.46)

and similarly for mode III:

$$G_\text{III} = (1+v)K_\text{III}^2/E.$$ (3.47)

(Note that an assumption implicit in the above relationships is that the crack extends *in its original plane*.) If the crack extends under combined-mode loading,

STRESSES AND DISPLACEMENTS IN CRACKED BODIES 39

then the total energy release rate (G_T) is given by:

$$G_T = G_I + G_{II} + G_{III} = \frac{1-v^2}{E}(K_I^2 + K_{II}^2 + K_{III}^2) \quad \text{plane strain} \quad (3.48)$$

3.8 Westergaard stress function

Referring to Section 1.7 the stresses and displacements in a body under general loading are given in terms of the complex stress functions, $\phi(z)$ and $\psi(z)$, as:

$$\sigma_x + \sigma_y = 4 \operatorname{Re}\{\phi'(z)\}$$
$$\sigma_y - \sigma_x + 2i\tau_{xy} = 2[\bar{z}\phi''(z) + \psi'(z)] \quad (3.49)$$

giving:

$$\sigma_y + i\tau_{xy} = 2\operatorname{Re}\{\phi'(z)\} + [\bar{z}\phi''(z) + \psi'(z)] \quad (3.50)$$

$$2\mu(u + iv) = \kappa\phi(z) - z\overline{\phi'(z)} - \overline{\psi(z)}. \quad (3.51)$$

In problems possessing x-axis symmetry for both loading and geometry, $\tau_{xy} = 0$ along $y = 0$. Hence, from (3.50):

$$\operatorname{Im}\{\bar{z}\phi''(z) + \psi'(z)\} = 0, \quad y = 0 \quad (3.52)$$

giving:

$$\{\bar{z}\phi''(z) + \psi'(z)\} = A, \quad y = 0 \quad (3.53)$$

where A is a real constant; thus for a traction-free crack:

$$2\operatorname{Re}\{\phi'(z)\} = -A, \quad y = 0, \quad -a \leqslant x \leqslant a. \quad (3.54)$$

The requirement of equation (3.53) is satisfied by:

$$\psi(z) = \phi(z) - z\phi'(z) + Az + B. \quad (3.55)$$

However, the newly introduced real constant, B, may be set to zero, since it only contributes a rigid body displacement. Substituting (3.55) the left-hand side of (3.52) becomes

$$\operatorname{Im}\{(\bar{z} - z)\phi''(z) + A\} \quad (3.56)$$

which is identically zero along $y = 0$. Now introduce a complex stress function $Z_I(z)$ such that:

$$Z_I(z) = 2\phi'(z) \quad (3.57)$$

$Z_I(z)$ is normally termed a Westergaard stress function [5]. The suffix 'I' indicates x-axis symmetry. (The reader should note that the original work [5] suggested stress functions which were not sufficiently general (see [6–8]). This omission is corrected by the inclusion of the real constant A above.)

Now rewrite the stress equations by substitution of (3.55) and (3.57) into (3.49), giving:

$$\sigma_x + \sigma_y = 2\,\text{Re}\{Z_\text{I}(z)\}$$

$$\sigma_y - \sigma_x + 2i\tau_{xy} = 2[(\bar{z} - z)Z_\text{I}'(z) + A].$$

Hence, taking real and imaginary parts:

$$\sigma_x = \text{Re}\{Z_\text{I}(z)\} - y\,\text{Im}\{Z_\text{I}'(z)\} - A$$
$$\sigma_y = \text{Re}\{Z_\text{I}(z)\} + y\,\text{Im}\{Z_\text{I}'(z)\} + A \tag{3.58}$$
$$\tau_{xy} = -y\,\text{Re}\{Z_\text{I}'(z)\}.$$

A similar procedure, detailed in [8], yields the displacement relationships:

$$\mu u = \left(\frac{\kappa - 1}{4}\right)\text{Re}\{\tilde{Z}_\text{I}(z)\} - \frac{y}{2}\text{Im}\{Z_\text{I}(z)\} - Ax$$

$$\mu v = \left(\frac{\kappa + 1}{4}\right)\text{Im}\{\tilde{Z}_\text{I}(z)\} - \frac{y}{2}\text{Re}\{Z_\text{I}(z)\} + Ay \tag{3.59}$$

where $2\phi(z) = \int Z_\text{I}(z)\,dz \equiv \tilde{Z}_\text{I}(z)$ and μ, κ are defined in Section 1.7.

Furthermore, by substituting from equation (3.57) into (3.13) we obtain stress intensity factor in terms of Z_I:

$$K_\text{I} = (2\pi)^{1/2} \lim_{z \to a} (z - a)^{1/2} Z_\text{I}(z). \tag{3.60}$$

Example 3.2

By using the expressions for $\phi(z)$ given in equations (3.5) derive the Westergaard stress function for a uniaxially loaded crack in an infinite sheet. Hence derive the stress intensity factor for a similar crack subjected to remote biaxial loading.

Solution: Differentiate equation (3.5) and substitute into equations (3.57) and (3.54) to obtain:

$$Z_\text{I}(z) = [\sigma z(z^2 - a^2)^{-1/2} - A], \quad A = \sigma/2. \tag{3.61}$$

For large $|z|$ equations (3.58) give:

$$\sigma_x = \sigma - A - A = 0, \quad A = \sigma/2$$
$$\sigma_y = \sigma - A + A = \sigma, \quad \text{all } A$$
$$\tau_{xy} = 0.$$

Thus, by manipulation of the real constant A we can decide the degree of biaxiality of the remote stress field. $A = \sigma/2$ produces a uniaxial field, $A = 0$ a

STRESSES AND DISPLACEMENTS IN CRACKED BODIES 41

biaxial field, giving:

$$Z_{\text{I}}(z) = [\sigma z(z^2 - a^2)^{-1/2}] \tag{3.62}$$

for remote biaxial stress. Solving for K_{I} by substituting from (3.62) into (3.60) we see that:

$$K_{\text{I}} = \sigma(\pi a)^{1/2} \tag{3.63}$$

whatever the value of A. Thus, the stress intensity factor is independent of the biaxiality of the stress field. However, the form of the non-singular terms in the stress-field equations will be changed [8].

Example 3.3

A Westergaard stress function is given as:

$$Z_{\text{I}} = \frac{Pa}{\pi z(z^2 - a^2)^{1/2}} \tag{3.64}$$

(a) What configuration and loading does this represent?
(b) What is the stress intensity factor?

Solution: (a) First, consider the remote loading. For $|z|$ very large, substitution of (3.64) into (3.58) shows that the remotely applied stresses are all zero:

$$\sigma_y = 0, \quad \sigma_x = 0, \quad \tau_{xy} = 0, \quad |z| \to \infty. \tag{3.65}$$

Now, along the x-axis:

$$Z_{\text{I}} = \frac{Pa}{\pi x(x^2 - a^2)^{1/2}} \tag{3.66}$$

but the term in parentheses is purely imaginary $a \leqslant x \leqslant a$, which appears to indicate a traction-free crack surface. However:

$$|\sigma_y| \to \infty \quad \text{as } x \to 0.$$

In order to evaluate the force acting over the upper portion of the crack surface we note that for equilibrium:

$$2\int_a^{+\infty} \sigma_y \, dx + V = 0 \tag{3.67}$$

where V is the vertical force per unit thickness on the upper crack surface at $x = 0$ (Fig. 3.7). Substituting (3.66) and (3.58) into (3.67) indicates that

$$V = -P$$

thus, the vertical force acting on the upper surface of the crack is of magnitude

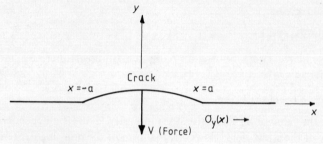

Fig. 3.7 *x-axis equilibrium.*

P, in the $+y$-direction. Symmetry of the Westergaard function indicates that a similar force acts on the lower crack surface in the $-y$-direction (Fig. 3.8).

(b) Substitute from (3.64) into (3.60) to obtain:

$$K_I = (2\pi)^{1/2} \lim_{z \to a} (z-a)^{1/2} \frac{P}{\pi z (z^2 - a^2)^{1/2}} \qquad (3.68)$$

which reduces immediately to:

$$K_I = \frac{P}{(\pi a)^{1/2}}. \qquad (3.69)$$

Evidently, the form of the singularity is the same as that noted previously for the case of remote tension. However, the magnitude of the singularity, the stress intensity factor, is different. The only difference between the point-loaded crack and the remotely loaded crack is the nature of the loading. Thus, the stress intensity factor is a function of the loading.

Now, a final example of the use of the Westergaard function, which will indicate the effect of the geometry on stress intensity, will be given.

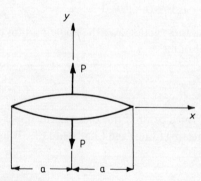

Fig. 3.8 *Wedge-loaded crack.*

STRESSES AND DISPLACEMENTS IN CRACKED BODIES

Example 3.4

(a) Show that the Westergaard stress function:

$$Z_I(z) = \left[\sigma \sin\left(\frac{\pi z}{2b}\right)\right]\left[\sin^2\left(\frac{\pi z}{2b}\right) - \sin^2\left(\frac{\pi a}{2b}\right)\right]^{1/2} \quad (3.70)$$

may be used to model the case of an array of collinear cracks subjected to remote loading.

(b) Derive the stress intensity factor for one crack in the array.

Solution: (a) For large $|z|$, $Z_I(z) \to \sigma$, thus from equations (3.58), the remote-loading conditions are:

$$\sigma_x = \sigma_y = \sigma, \quad \tau_{xy} = 0$$

which is a state of equal biaxial tension. Along the x-axis, $Z_I(z)$ is purely imaginary, $|x - 2nb| \leq a$, where $n = 1, 2, 3 \ldots$. Hence, from (3.58)

$$\sigma_y = 0, \quad |x - 2nb| \leq a$$

and the cracks are periodic, of period $2b$ in x (Fig. 3.9).

Thus, $Z_I(z)$ defined in equation (3.70) is the Westergaard function for an array of cracks in a biaxially loaded sheet.

(b) We shall derive K_I via equation (3.12), namely

$$K_I = (2\pi)^{1/2} \lim_{z \to a} (z-a)^{1/2} \sigma_y. \quad (3.71)$$

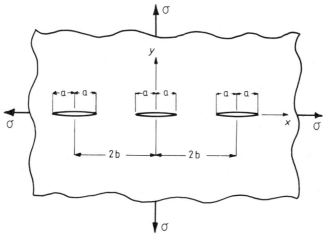

Fig. 3.9 *Periodic array of cracks under biaxial stress.*

44 THE MECHANICS OF FRACTURE AND FATIGUE

By substitution from (3.70) into (3.58) we obtain:

$$\sigma_y = \text{Re}\left\{\left[\sigma \sin\left(\frac{\pi x}{2b}\right)\right]\left[\sin^2\left(\frac{\pi x}{2b}\right) - \sin^2\left(\frac{\pi a}{2b}\right)\right]^{-1/2}\right\}. \quad (3.72)$$

Note that:

$$(2\pi)^{1/2} \lim_{x \to a} (x-a)^{1/2}\, \sigma_y = \lim_{x \to a} (8b)^{1/2} \sin\left(\frac{\pi x}{4b} - \frac{\pi a}{4b}\right)^{1/2} \sigma_y \quad (3.73)$$

which after substituting for σ_y from (3.72) yields the solution for stress intensity:

$$K_\text{I} = \sigma(\pi a)^{1/2}\left[\frac{2b}{\pi a} \tan\left(\frac{\pi a}{2b}\right)\right]^{1/2}. \quad (3.74)$$

(Note that the above solution reproduces the single crack solution: equation (3.63) as $b/a \to \infty$.) This example shows that the magnitude of K_I depends on the geometry of the body, whilst Example 3.3 indicates that K_I is also a function of loading.

3.9 The configuration correction factor

It is common practice to use the infinite-sheet solution to non-dimensionalize K_I results, hence producing a factor Q, where:

$$K_\text{I} = Q\sigma(\pi a)^{1/2}. \quad (3.75)$$

In Example 3.4, for instance:

$$Q = \left[\frac{2b}{\pi a} \tan\left(\frac{\pi a}{2b}\right)\right]^{1/2}. \quad (3.76)$$

Q is frequently termed a geometrical factor. This is strictly incorrect, since Q is a function of both geometry and loading. Herein Q will be referred to as a *configuration correction factor*, where configuration implies a combination of geometry and loading. The importance of separating out geometry and loading will become clear in Chapter 4.

3.10 The Williams stress function

A series expansion of a generalized form of the complex stress functions for a cracked body [9], or equivalently, a special case of the Airy stress function for an infinite sector satisfying zero normal and shear tractions along the line of a semi-infinite crack (Fig. 3.10) is termed the Williams stress function [10], F_W. The Williams stress function may be written as:

$$F_\text{W} = F_\text{W}^\text{e} + F_\text{W}^\text{o} \quad (3.77)$$

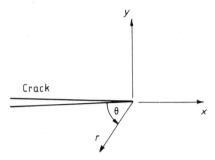

Fig. 3.10 *Stress-free, semi-infinite crack.*

where F_W^e and F_W^o are even and odd terms respectively, given by:

$$F_W^e = \sum_{n=1}^{\infty} \left[(-1)^{n-1} A_{2n-1} r^{n+\frac{1}{2}} \left(-\cos[(n-\tfrac{3}{2})\theta] + \frac{2n-3}{2n+1} \cos[(n+\tfrac{1}{2})\theta] \right) \right.$$
$$\left. + (-1)^n A_{2n} r^{n+1} \left\{ -\cos[(n-1)\theta] + \cos[(n+1)\theta] \right\} \right]$$

$$F_W^o = \sum_{n=1}^{\infty} \left[(-1)^{n-1} B_{2n-1} r^{n+\frac{1}{2}} \left\{ \sin[(n-\tfrac{3}{2})\theta] - \sin[(n+\tfrac{1}{2})\theta] \right\} \right. \quad (3.78)$$
$$\left. + (-1)^n B_{2n} r^{n+1} \left(-\sin[(n-1)\theta] + \frac{n-1}{n+1} \sin[(n+1)\theta] \right) \right].$$

(Note that equations (3.78) above are corrected in accordance with [11].) The A's and B's are real coefficients having the dimensions of stress.

Conversion of the Airy relationships of Section 1.6 into polar coordinates and stresses gives [12]:

$$\sigma_r = \frac{1}{r} \frac{\partial F}{\partial r} + \frac{1}{r^2} \frac{\partial^2 F}{\partial \theta^2},$$

$$\sigma_\theta = \frac{\partial^2 F}{\partial r^2}, \quad (3.79)$$

$$\tau_{r\theta} = -\frac{\partial}{\partial r} \frac{1}{r} \frac{\partial F}{\partial \theta}.$$

In the case of the Williams stress function (equations (3.77) and (3.78)) use of the transformation relationships (Section 1.2.2), together with the stress intensity definitions of equations (3.38) and (3.40), leads to:

$$K_I = -A_1 (2\pi)^{1/2}, \quad K_{II} = B_1 (2\pi)^{1/2}. \quad (3.80)$$

Problems

3.1. The stress function $\phi'(z)$ for the configuration illustrated in Fig. 3.11 is given as:

$$\phi'(z) = \frac{P}{2\pi}\left(1 - \frac{y_0}{2(1-v)}\frac{\partial}{\partial y_0}\right)[f(z,y,a) - f(z,y_0,a)]$$

$$f(z,y_0,a) = \frac{(a^2 + y_0^2)^{1/2}}{z^2 + y_0^2}\frac{1}{(1 - a^2/z^2)^{1/2}}.$$

Determine the stress intensity factor.

Answer:

$$K_I = P\left(\frac{a}{\pi}\right)^{1/2}\left(\frac{1}{(a^2 + y_0^2)^{1/2}} + \frac{y_0^2}{2(1-v)(a^2 + y_0^2)^{3/2}}\right)$$

$$K_{II} = K_{III} = 0.$$

3.2 By resolution of remote loading into direct and shear components determine stress intensity factors for the configuration shown in Fig. 3.12.

Answer:

$$K_I = (\sigma \sin^2 \beta)(\pi a)^{1/2}$$

$$K_{II} = (\sigma \sin \beta \cos \beta)(\pi a)^{1/2}$$

$$K_{III} = 0.$$

3.3 By adopting an approach similar to that employed in section 3.8:

(a) Derive expressions for the stresses in the case of crack-line anti-symmetry

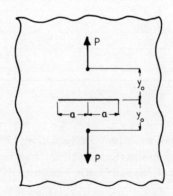

Fig. 3.11 *Opposing point forces, cracked infinite sheet.*

STRESSES AND DISPLACEMENTS IN CRACKED BODIES

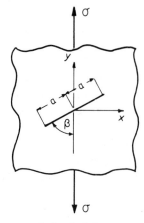

Fig. 3.12 Angled crack in remotely loaded sheet.

($\sigma_y = 0$ along $y = 0$), in terms of the Westergaard stress function $Z_{II}(z) = 2\phi'(z)$.

(b) Obtain expressions for the crack shape in terms of $Z_{II}(z)$.

(c) Find stress intensity factor K_{II} in terms of $Z_{II}(z)$.

Answer:

(a) $\sigma_x = 2 \operatorname{Re}\{Z_{II}(z)\} - y \operatorname{Im}\{Z'_{II}(z)\}$

$\sigma_y = y \operatorname{Im}\{Z'_{II}(z)\}$

$\tau_{xy} = -\operatorname{Im}\{Z_{II}(z)\} - y \operatorname{Re}\{Z'_{II}(z)\} + B$

where B is a real constant.

(b) $2\mu u = \left(\dfrac{\kappa + 1}{2}\right) \operatorname{Re}\{\tilde{Z}_{II}(z)\} - y \operatorname{Im}\{Z_{II}(z)\} + By$

$2\mu v = \left(\dfrac{\kappa - 1}{2}\right) \operatorname{Im}\{\tilde{Z}_{II}(z)\} - y \operatorname{Re}\{Z_{II}(z)\} + Bx$

(c) $K = \lim\limits_{z \to a} (2\pi)^{1/2} (z - a)^{1/2} Z_{II}(z)$.

(The full derivation is given in [8].)

3.4 Use the stress relationship given in equation (3.79) to show that the Williams stress function (equations (3.77) and (3.78)) defines a stress-free, semi-infinite crack.

References

1. Sneddon I. N. (1946), 'The distribution of stress in the neighbourhood of a crack in an elastic solid', *Proc. R. Soc., Lond.* A, **187**, 229–60.
2. Irwin G. R. (1957), 'Analysis of stresses and strains near the end of a crack traversing a plate', *Trans. ASME, J. Appl. Mech.*, **24**, 361–4.
3. Sih G. C. and Liebowitz H. (1968), 'Mathematical theories of brittle fracture' in *Fracture – An Advanced Treatise*, Vol. II, Ed. H. Liebowitz, Academic Press, New York.
4. Paris P. C. and Sih G. C. (1965), 'Stress analysis of cracks' in *Fracture Toughness Testing and its Applications*, ASTM STP 381, 30–77.
5. Westergaard H. M. (1939), 'Bearing pressures and cracks', *J. Appl. Mech.*, **6**, 49–53.
6. Sih G. C. (1966), 'On the Westergaard method of crack analysis', *Int. J. Frac. Mech.*, **2**, 628–31.
7. Eftis J. and Liebowitz H. (1972), 'On the modified Westergaard equations for certain plane crack problems', *Int. J. Frac. Mech.*, **8**, 383–91.
8. Eftis, J., Subramonian N. and Liebowitz H. (1977), 'Crack border stress and displacement equations revisited' *Engng Frac. Mech.*, **9**, 189–210.
9. Rice J. R. (1968), 'Mathematical analysis in the mechanics of fracture' in *Fracture – An Advanced Treatise*, Vol. II, Ed. H. Liebowitz. Academic Press, New York.
10. Williams M. L. (1957), 'On the stress distribution at the base of a stationary crack', *J. Appl. Mech.*, **24**, 109–14.
11. Cartwright D. J. and Rooke D. P. (1975), 'Evaluation of stress intensity factors', *J. Strain Analysis*, **10**, No. 4, 217–24.
12. Timoshenko S. P. and Goodier J. N. (1970), *Elements of Elasticity*, 3rd edn, McGraw-Hill, New York.

4 Determination of Stress Intensity Factors

4.1 Introduction

The stress intensity factor is of fundamental importance in the prediction of brittle failure using linear elastic fracture mechanics (LEFM) principles. It is a function of both the cracked geometry and the associated loading. It is common practice to present K solutions in dimensionless form, normalized with respect to an appropriate infinite-sheet solution.

The problem facing a designer is to strike a balance between time, cost and accuracy in selecting a suitable method for determining stress intensities. In a relatively short chapter of an introductory text it is clearly not feasible to cover each method in detail, nor is it desirable if the reader is seeking a general appreciation. Each solution technique includes the following details:

(a) A description of the method. This is normally fairly brief if other sources for introductory reference are available. Areas which are not covered elsewhere at a suitable introductory level include weight functions and boundary collocation.
(b) Accuracy of the method.
(c) Time required to obtain a solution.
(d) Possible future developments.

4.2 Analytical

Analytical solutions are those which satisfy all the boundary conditions exactly. Only relatively simple geometries are soluble by analytical methods, usually this implies a body of infinite extent.

We have already met one analytic technique in the previous chapter, namely the use of Westergaard stress functions. Several workers have employed Westergaard stress functions [1–3], and recent papers have noted omissions in the original formulation [4, 5], and the equivalence of Muskhelishvili's complex stress function method [6].

The advantage of the complex stress function approach is that it extends the possible range of problems by allowing conformal mapping of (say) a circular hole into a crack subjected to general loading.

For mixed-mode problems, the general form of equation (3.13) for a crack orientated along the x-axis with its tip located at z_1, is:

$$K = K_\mathrm{I} - iK_\mathrm{II} = 2(2\pi)^{1/2} \lim_{z \to z_1} (z - z_1)^{1/2} \phi'(z). \tag{4.1}$$

By employing a mapping function $z = \omega(\zeta)$, following the approach described in Section 1.8, we obtain:

$$K = 2(2\pi)^{1/2} \lim_{\zeta \to \zeta_1} [\omega(\zeta) - \omega(\zeta_1)]^{1/2} \frac{\phi'(\zeta)}{\omega'(\zeta)}. \tag{4.2}$$

This expression is considerably simplified by the mapping which takes a crack of length $2a$ on to a circle of unit radius:

$$z = \omega(\zeta) = \frac{a}{2}\left(\zeta + \frac{1}{\zeta}\right) \tag{4.3}$$

giving:

$$K = 2\left(\frac{\pi}{a}\right)^{1/2} \phi'(1). \tag{4.4}$$

For instance, a knowledge of $\phi(\zeta)$ for an inclined point load acting on the unit circle in the ζ-plane, may be used to derive K for the equivalent problem in the

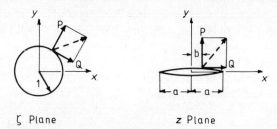

ζ Plane z Plane

Fig. 4.1 *Point force, infinite sheet (after [7]).*

z-plane containing a crack (Fig. 4.1). The resulting stress intensity factors are given by [7]:

$$\begin{aligned}
K_\mathrm{I} &= \frac{P}{2(\pi a)^{1/2}}\left(\frac{a+b}{a-b}\right)^{1/2} + \frac{Q}{2(\pi a)^{1/2}}\left(\frac{\kappa-1}{\kappa+1}\right) \\
K_\mathrm{II} &= -\frac{P}{2(\pi a)^{1/2}}\left(\frac{\kappa-1}{\kappa+1}\right) + \frac{Q}{2(\pi a)^{1/2}}\left(\frac{a+b}{a-b}\right).
\end{aligned} \tag{4.5}$$

Further examples of solutions obtained by this method are outlined in [7].

4.3 Green's functions

Results for a generally positioned point force may provide Green's functions to solve the problem of the same geometry under a general, distributed loading. For instance, the results quoted in equation (4.5) give directly the following expressions for a single crack subjected to distributed tractions:

$$K_\mathrm{I} = \frac{1}{(\pi a)^{1/2}} \int_{-a}^{a} \sigma_y(x, 0) \left(\frac{a + x}{a - x}\right)^{1/2} dx$$

$$K_\mathrm{II} = \frac{1}{(\pi a)^{1/2}} \int_{-a}^{a} \tau_{xy}(x, 0) \left(\frac{a + x}{a - x}\right)^{1/2} dx$$

(4.6)

where σ_y and τ_{xy} are the direct and shear stresses acting along the crack line in the absence of the crack.

The above expression was derived from the analytic Green's function, equation (4.5). A Green's function contains information on the effect on K of a generally positioned point force, and this information may be obtained by any suitable method, the accuracy associated with the final solution being dependent upon the accuracy of the original point force solution, and any subsequent numerical manipulation, e.g. curve fitting. In general the expression for, say K_I, in terms of the Green's function $G(x)$, may be written:

$$K_\mathrm{I} = \frac{1}{(\pi a)^{1/2}} \int_{a}^{\cdot} p(x)\, G(x)\, dx \qquad (4.7)$$

where $p(x)$ is the pressure acting normal to the crack surface.

It may be demonstrated [8], that certain simple methods of determining K can be expressed in terms of approximate Green's functions; in particular the approximation methods based on maximum stress at the notch root, stress at crack tip location in the unflawed structure, and average stress over the crack line. Of these three methods, the second gives the best approximation because of the extreme dependence of K on the crack tip stresses. Full details, together with a summary and appropriate solution techniques, are given in [8].

The time taken to obtain additional solutions is short (two or three days), provided suitable Green's functions are available, and crack-line loading is known.

4.4 Weight function techniques

Bueckner [9] and Rice [10] have demonstrated that a particular function, normally termed the Bueckner weight function, is a property of a cracked geometry and is independent of the loading. The weight function may be employed in the derivation of additional stress intensity factor solutions provided details of crack-line loading are available. A weight function may be thought of as a form of the Green's function for a cracked body.

This section is intended to give the reader an understanding of the meaning and applications of weight function techniques. A proof of the uniqueness of the weight function is included in an appendix at the end of the chapter.

4.4.1 *Derivation*

The relationship between stress intensity factor, K_I, and strain energy release rate per unit crack extension, G, has been derived, see equations (3.14) and (3.15):

$$K_I = (HG)^{1/2} \tag{4.8}$$

where

$$H = \begin{cases} E, & \text{plane stress} \\ E/(1-v^2), & \text{plane strain} \end{cases}$$

Consider a configuration, having x-axis symmetry, containing a crack of length a (Fig. 4.2) subjected to a system of loads which have the same symmetry. From Section 3.6 the difference between the strain energy of this cracked structure, and that of the same structure in the absence of the crack is given by:

$$U = \tfrac{1}{2} \int_a p(x)\, v(x, a)\, dx \tag{4.9}$$

where $p(x)$ is the stress distribution along the x-axis in the uncracked structure,

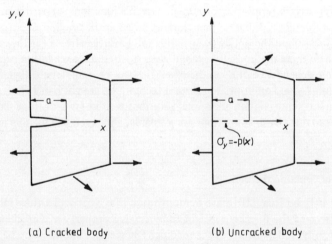

(a) Cracked body   (b) Uncracked body

Fig. 4.2 *Cracked and uncracked bodies subjected to identical loading.*

DETERMINATION OF STRESS INTENSITY FACTORS 53

and v is the displacement in the y-direction of a point at position x on the crack, caused by the action of the loading system.

From the definition of G (note G is a per tip value):

$$G = \frac{\partial U}{\partial a} = \tfrac{1}{2} \int_a p(x) \frac{\partial v(x, a)}{\partial a} dx. \qquad (4.10)$$

Note that $p(x)$ is not a function of a in the case considered.

Substitute from (4.8) into (4.10) to obtain:

$$K_I = \int_a p(x) m(x, a) dx \qquad (4.11)$$

where

$$m(x, a) = \frac{H}{2K_I} \frac{\partial v(x, a)}{\partial a} \qquad (4.12)$$

$m(x, a)$ is the Bueckner weight function. It can be demonstrated that this weight function is *unique to the given geometry* and is *independent of the loading* from which it was derived. (See appendix at the end of this chapter for proof.)

Accepting the uniqueness of the weight function, $m(x, a)$, it is evident that the weight function could have been derived from any loading applied to the geometry, and from its associated crack shape. Hence, additional stress intensity factor solutions, K_I^* under crack-line loading $p^*(x)$ may be obtained from

$$K_I^* = \int_a p^*(x) m(x, a) dx. \qquad (4.13)$$

(Caution: the loading $p(x)$ used in the derivation of $m(x, a)$ must not have more symmetries than the new loading, $p^*(x)$. Thus, if $m(x, a)$ is derived from a $p(x)$ having biaxial symmetry, it is only applicable to a $p^*(x)$ having biaxial symmetry.)

We have only considered the possibility of a loaded crack surface. This is not as restricting as it appears at first sight, since superposition techniques allow the 'transfer' of loading to the crack line (see Example 3.1). However, the energy changes associated with incremental crack extension concern displacements on all loaded boundaries. Furthermore, an additional term is required in (4.11) if body forces are present. The inclusion of these two caveats serves to generalize equation (4.11) to:

$$K_I^* = \int_\Gamma t \cdot h \, d\Gamma + \int_A f \cdot h \, dA \qquad (4.14)$$

where

$$h = h(x, y, a) = \frac{H}{2K} \frac{\partial u}{\partial a} \qquad (4.15)$$

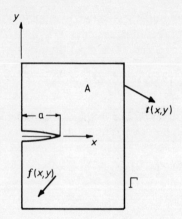

Fig. 4.3 *Boundary and body force loadings (after [10]).*

u is the displacement field within the body, the stress vector t acts on the boundary Γ, and the body force f within the region A enclosed by Γ (Fig. 4.3).

Knowing K_1^* it is possible to return to equations (4.12) or (4.15) and integrate with respect to a in order to solve for the new displacement field v^*, provided it is known for one value of crack length, a [10].

Example 4.1

The stress intensity factor and associated crack shape for a single crack in a remotely loaded infinite sheet (Fig. 4.4(a)) are given in equations (3.11) and (3.29) respectively.

(a) Derive the weight function for this geometry.

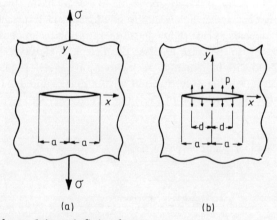

Fig. 4.4 *Single crack in an infinite sheet.*

DETERMINATION OF STRESS INTENSITY FACTORS

(b) Calculate the stress intensity factor for a centrally located band of pressure applied to the crack surface (Fig. 4.4(b)).
(c) Calculate the stress intensity factor for a central point force applied to the crack surface.
(d) Derive an expression for the crack shape with point force loading.

Solution: (a) From (3.11)

$$K_I = \sigma(\pi a)^{1/2}. \tag{4.16}$$

The crack shape is given by (3.29) as:

$$v(x, a) = \left(\frac{\kappa + 1}{4\mu}\right) \sigma(a^2 - x^2)^{1/2}, \quad -a \leq x \leq a. \tag{4.17}$$

Thus

$$\frac{\partial v}{\partial a} = \left(\frac{\kappa + 1}{4\mu}\right) \sigma a (a^2 - x^2)^{-1/2}. \tag{4.18}$$

Substituting (4.16) and (4.18) into (4.12) noting that $H((\kappa + 1)/4\mu) = 2$ for plane stress and plane strain, we obtain the weight function

$$m(x, a) = \left(\frac{a}{\pi}\right)^{1/2} (a^2 - x^2)^{-1/2} \tag{4.19}$$

(b) The required stress intensity factor K_I^* is obtained by substituting from (4.19) into (4.13), giving:

$$K_I^* = \left(\frac{a}{\pi}\right)^{1/2} \int_a \frac{p(x)}{(a^2 - x^2)^{1/2}} dx \tag{4.20}$$

Now, for the right-hand crack tip (say), the integral is taken over the upper and lower, right-hand crack surfaces, so that (4.20) becomes:

$$K_I^* = \left(\frac{a}{\pi}\right)^{1/2} \cdot 2 \int_0^a \frac{p(x)}{(a^2 - x^2)^{1/2}} dx. \tag{4.21}$$

Thus, if $p(x)$ is a band of constant pressure p, acting over upper and lower crack surfaces from $x = -d$ to $x = +d$:

$$K_I^* = \left(\frac{a}{\pi}\right)^{1/2} \cdot 2p \int_0^d \frac{1}{(a^2 - x^2)^{1/2}} dx \tag{4.22}$$

which yields the required stress intensity:

$$K_I^* = 2\left(\frac{a}{\pi}\right)^{1/2} p \arcsin\left(\frac{d}{a}\right) \tag{4.23}$$

(c) In the case of equal and opposite point forces P, let:

$$P = 2pd \tag{4.24}$$

and the required result is given by:

$$K_I^* = \lim_{d \to 0} 2\left(\frac{a}{\pi}\right)^{1/2} \frac{P}{2d} \arcsin\left(\frac{d}{a}\right) \tag{4.25}$$

thus

$$K_I^* = P/(\pi a)^{1/2} \tag{4.26}$$

which conforms with the solution obtained in Example 3.3.

(d) We now know the stress intensity factor for the point force problem and the associated weight function. Re-arranging equation (4.12):

$$\frac{\partial v}{\partial a} = \frac{2K}{H} m(x, a) \tag{4.27}$$

which gives, after substituting from (4.19) and (4.26):

$$v = \frac{2P}{\pi H} \int \frac{1}{(a^2 - x^2)^{1/2}} \, da. \tag{4.28}$$

Note that when $x = a$, $v = 0$, thus,

$$v = \frac{2P}{\pi H} \ln\left(\frac{a + (a^2 - x^2)^{1/2}}{x}\right). \tag{4.29}$$

This result may be confirmed by the use of Westergaard stress functions, by substituting from (3.64) into (3.59).

Example 4.2

The stress intensity factor for an array of equal length cracks loaded remotely by a stress σ is given in equation (3.74) as:

$$K = \sigma\left[2b \tan\left(\frac{\pi a}{2b}\right)\right]^{1/2} \tag{4.30}$$

where $2a$ is the crack length, and $2b$ the distance between crack centres. The crack shape may be obtained from equations (3.70) and (3.59), and is, for plane strain [11]:

$$v(x, a) = \frac{2b}{\pi a} \epsilon^{\infty} \left(\ln\left\{ \cos\left(\frac{\pi x}{2b}\right) + \left[\cos^2\left(\frac{\pi x}{2b}\right) - \cos^2\left(\frac{\pi a}{2b}\right)\right]^{1/2} \right\} \right.$$
$$\left. - \ln \cos\left(\frac{\pi a}{2b}\right) \right), \tag{4.31}$$

DETERMINATION OF STRESS INTENSITY FACTORS

where

$$\epsilon^\infty = \frac{2(1-v^2)\sigma a}{E}. \quad (4.32)$$

(a) Derive the weight function for this geometry.
(b) Calculate the stress intensity factor for a centrally located band of constant pressure applied to each of the crack surfaces (Fig. 4.5).
(c) Calculate the stress intensity factor for a central point force applied to the crack surface.
(d) Suggest and apply a superposition technique to obtain stress intensity factors for an array of rivet-loaded cracks in a sheet.

Solution: (a) For the geometry under consideration, differentiation of equation (4.31) yields:

$$\frac{\partial v(x,a)}{\partial a} = \frac{4b(1-v^2)}{\pi E}$$

$$\times \sigma \left(\frac{\frac{\pi}{2b}\cos\left(\frac{\pi a}{2b}\right)\sin\left(\frac{\pi a}{2b}\right)\left[\cos^2\left(\frac{\pi x}{2b}\right) - \cos^2\left(\frac{\pi a}{2b}\right)\right]^{-1/2}}{\cos\left(\frac{\pi x}{2b}\right) + \left[\cos^2\left(\frac{\pi x}{2b}\right) - \cos^2\left(\frac{\pi a}{2b}\right)\right]^{1/2}} + \frac{\pi}{2b}\tan\left(\frac{\pi a}{2b}\right) \right).$$

(4.33)

Rewriting $\cos^2(\pi x/2b) - \cos^2(\pi a/2b)$ as $\sin^2(\pi a/2b) - \sin^2(\pi x/2b)$ and making simplifications, we obtain:

$$\frac{\partial v(x,a)}{\partial a} = \frac{2(1-v^2)}{E} \sigma \tan\left(\frac{\pi a}{2b}\right) \frac{\cos\left(\frac{\pi x}{2b}\right)}{\left[\sin^2\left(\frac{\pi a}{2b}\right) - \sin^2\left(\frac{\pi x}{2b}\right)\right]^{1/2}}. \quad (4.34)$$

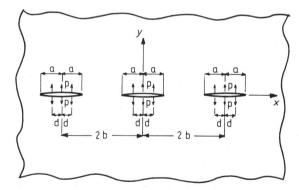

Fig. 4.5 *Array of cracks, each loaded by band of pressure.*

Substitute for K and $\partial v(x,a)/\partial a$ from (4.30) and (4.34) into (4.12) to obtain the weight function:

$$m(x, a) = \frac{1}{2b}\left[2b \tan\left(\frac{\pi a}{2b}\right)\right]^{1/2} \frac{\cos\left(\frac{\pi x}{2b}\right)}{\left[\sin^2\left(\frac{\pi a}{2b}\right) - \sin^2\left(\frac{\pi x}{2b}\right)\right]^{1/2}}.$$

(4.35)

(b) Substitute equation (4.35) into (4.13) to obtain the stress intensity factor

$$K_I^* = \frac{1}{2b}\left[2b \tan\left(\frac{\pi a}{2b}\right)\right]^{1/2} \cdot 2\int_0^a p^*(x) \frac{\cos\left(\frac{\pi x}{2b}\right)}{\left[\sin^2\left(\frac{\pi a}{2b}\right) - \sin^2\left(\frac{\pi x}{2b}\right)\right]^{1/2}} dx.$$

(4.36)

To test this solution as the crack spacing increases, note that

$$\lim_{b \to \infty} K_I^* = 2\left(\frac{a}{\pi}\right)^{1/2} \int_0^a \frac{p^*(x)}{(a^2 - x^2)^{1/2}} dx,$$

(4.37)

which conforms with the solution for a single symmetrically loaded crack in an infinite sheet, obtained in the previous example.

For the case of a band of constant pressure, p, acting over upper and lower crack faces from $-d$ to d, for any d between 0 and a equation (4.36) becomes

$$K_I^* = p\left(2b \tan\frac{\pi a}{2b}\right)^{1/2} \cdot \frac{1}{2b} \cdot 2\int_0^d \frac{\cos\left(\frac{\pi x}{2b}\right)}{\left[\sin^2\left(\frac{\pi a}{2b}\right) - \sin^2\left(\frac{\pi x}{2b}\right)\right]^{1/2}} dx. \quad (4.38)$$

By making the substitution

$$\gamma = \frac{\sin(\pi x/2b)}{\sin(\pi a/2b)}, \quad d\gamma = \frac{\pi}{2b} \frac{\cos(\pi x/2b)}{\sin(\pi a/2b)} dx \quad (4.39)$$

equation (4.38) can be integrated immediately to give

$$K_I^* = p\left[2b \tan\left(\frac{\pi a}{2b}\right)\right]^{1/2} \left(\frac{2}{\pi}\right) \arcsin\left(\frac{\sin(\pi d/2b)}{\sin(\pi a/2b)}\right). \quad (4.40)$$

(c) $d \to 0$, i.e. central point loading for an infinite array of cracks. In this case (4.40) gives:

$$\lim_{d \to 0} \frac{K_I^*}{2pd} = \left[b \sin\left(\frac{\pi a}{b}\right)\right]^{-1/2} \quad (4.41)$$

DETERMINATION OF STRESS INTENSITY FACTORS

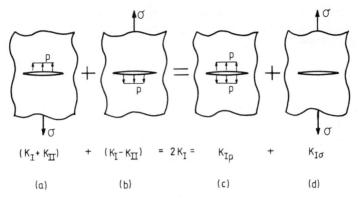

Fig. 4.6 *Superposition technique to model rivet-loading.*

(d) Application to rivet-loaded configurations. A simple model to represent an array of rivet-loaded cracks is illustrated in Fig. 4.6(a). The opening-mode stress intensity factor, K_I, for this configuration may be obtained by the superposition technique illustrated in Fig. 4.6. The K_I contributions arising from pressure on crack faces (K_{Ip}) and from remote uniaxial stress ($K_{I\sigma}$) are separated. Note that the superposition allows K_I to be obtained, but does not yield the sliding-mode stress intensity factor K_{II} [12]. Fig. 4.7 shows a plot of the superposed,

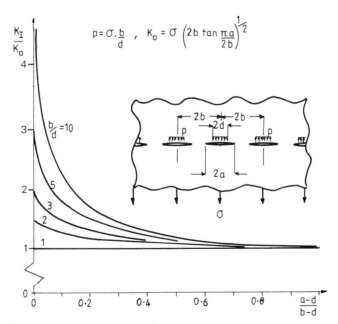

Fig. 4.7 *K-values for rivet-load model.*

dimensionless K_I solution for various b/d ratios. This presentation keeps the ratio of pressure band dimension to crack spacing constant, in order to model a rivet-loaded configuration.

Example 4.3

The weight function for a finite width, edge-cracked, infinite strip (Fig. 4.8(a)) is given as [13]:

$$m(x, a) = [2\pi(a - x)]^{-1/2} \left[1 + m_1\left(\frac{a-x}{a}\right) + m_2\left(\frac{a-x}{a}\right)^2\right] \quad (4.42)$$

m_1 and m_2 are functions of the ratio of crack depth to strip width, a/W, and are given as (for $0 \leq a/W \leq 0.5$):

$$m_1 = A_1 + B_1\left(\frac{a}{W}\right)^2 + C_1\left(\frac{a}{W}\right)^6, \quad m_2 = A_2 + B_2\left(\frac{a}{W}\right)^2 + C_2\left(\frac{a}{W}\right)^6 \quad (4.43)$$

where

$$A_1 = 0.6147, \quad B_1 = 17.1844, \quad C_1 = 8.7822$$
$$A_2 = 0.2502, \quad B_2 = 3.2899, \quad C_2 = 70.0444.$$

(a) Calculate the stress intensity for a crack of length $W/2$ in a strip subjected to a remote stress σ.
(b) Calculate the stress intensity factor for the same geometry subjected to a remote bending moment M.
(c) Calculate the crack mouth displacement for cases (a) and (b).

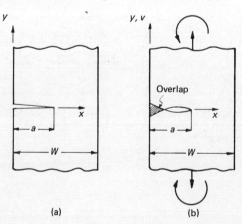

Fig. 4.8 *Edge-cracked, infinite strip:* (a) *geometry;* (b) *physically unacceptable overlapping.*

DETERMINATION OF STRESS INTENSITY FACTORS

(d) By choosing a specific combination of loading systems (a) and (b) show that it is possible to obtain a positive stress intensity with apparent overlapping of the crack surfaces.

Solution: (a) Substituting from (4.42) into (4.13), recognizing that the crack-line stresses in the absence of the crack are simply constant stress, σ:

$$K_I^* = 2\sigma \int_0^a [2\pi(a-x)]^{-1/2} \left[1 + m_1 \left(\frac{a-x}{a}\right) + m_2 \left(\frac{a-x}{a}\right)^2\right] dx \quad (4.44)$$

Evaluate m_1 and m_2 for $a/W = 0.5$, from (4.43):

$$m_1 = 5.048, \quad m_2 = 2.1669$$

which gives

$$K_I^* = 4.9724 \, \sigma a^{1/2} \quad (4.45)$$

whilst

$$Q = K_I^*/\sigma(\pi a)^{1/2} = 2.805$$

(b) For the case of pure bending, the crack-line stresses for a beam of unit thick thickness, are given by:

$$\sigma = \alpha(a-x) \quad (4.46)$$

where

$$\alpha = 12 \, M/W^3 \quad (4.47)$$

whilst the maximum stress, σ_{max} is:

$$\sigma_{max} = \alpha a. \quad (4.48)$$

Substituting from (4.42) and (4.46) into (4.13) we obtain:

$$K_I^* = \frac{2\alpha}{(2\pi)^{1/2}} \int_0^a (a-x)^{1/2} \left[1 + m_1 \left(\frac{a-x}{a}\right) + m_2 \left(\frac{a-x}{a}\right)^2\right] dx. \quad (4.49)$$

which yields:

$$K_I^* = 2.637 \, \alpha a^{3/2} \quad (4.50)$$

Now, non-dimensionalizing:

$$Q = \frac{K_I^*}{\sigma_{max}(\pi a)^{1/2}} = 1.488.$$

(c) From equation (4.12) we note:

$$\frac{\partial v}{\partial a} = \frac{2K}{H} m(x, a). \quad (4.51)$$

Since we know that $v = 0$ at $x = 0$, this expression may be integrated with respect to a to obtain the crack displacement, v:

$$v = \frac{2K}{H} m(x, a)\, da \qquad (4.52)$$

Note that in general K is a function of a, and the integration procedure may become unwieldy. However, we can obtain directly the crack mouth displacements, v_0, in the form:

$$v_0 = \frac{2\sigma a}{H} V(a/W) \qquad (4.53)$$

where $V(a/W)$ is given in [14]. For remote tension, case (a), we obtain:

$$v_0 = \frac{9.85}{H} \sigma a \qquad (4.54)$$

The equivalent result for pure bending, case (b) is:

$$v_0 = \frac{6.38}{H} \sigma_{max} a. \qquad (4.55)$$

(d) Consider the case in which remote tension σ is applied at the same time as pure bending, the latter producing a σ_{max} equal to -1.7σ. Superposition of these two loading systems produces a total stress intensity, K_S, given by:

$$K_S = \sigma(\pi a)^{1/2} [2.805 - 1.7(1.488)] = +0.275\, \sigma(\pi a)^{1/2}. \qquad (4.56)$$

In similar vein, superposition of the crack mouth displacements gives:

$$v_S = \frac{\sigma a}{H} [9.85 - 1.7(6.38)] = -1.00 \frac{\sigma a}{H} \qquad (4.57)$$

The shape of the crack surfaces is shown schematically in Fig. 4.8(b).

Evidently, it is possible to obtain a positive K-value, associated with a physically unacceptable overlapping of the crack surfaces some distance from the crack tip. This point is of some importance. It is obviously necessary to check that 'overlapping' of the crack surfaces has not occurred, even though the K-value may be positive. The question of overlapping is considered in [15].

Whilst space does not permit a more detailed examination, it should be clear to the reader that a relatively straightforward procedure, based on the weight function, should allow the inclusion of additional crack-line stresses which will eliminate overlapping.

4.4.2 *Additional remarks*

The weight function technique has been employed by workers to obtain approximate solutions [16, 17]. One particularly promising approach is the determination of weight function data from numerical results in order that designers may apply their own loading to a given geometry [18, 19]. Weight functions are also applicable to three-dimensional problems [20]. Accurate stress intensity factor solutions normally involve a significant effort. Crack shape data, suitable for the derivation of weight functions, is not normally extracted, although the additional effort would be minimal. Quite simply, it is a waste of time and money if weight function data are not derived as a by-product of a stress intensity factor solution. Nevertheless, it is not yet standard procedure to publish full weight function details, possibly because of the ease with which additional solutions, and hence additional papers, may be generated.

Given that stress intensity factor solutions are available, a technique proposed by Petroski and Achenbach [21] may be employed to derive weight functions. Briefly, this consists of using equation (4.11) in which $K_I(a)$ and the crack-line loading $p(x)$ are known from the original solution, in order to solve for the weight function $m(x, a)$. The functional dependence of v on both x and a is not known, however [21] proposes a particular dependence on x, based on correct behaviour and simplicity in use, in order to obtain solutions.

4.5 Boundary collocation

Boundary collocation is a numerical technique used to obtain solutions to various types of boundary value problems [22–25]. In the case of linear, two-dimensional elasticity it consists of taking an exact series solution to the governing biharmonic differential equations, in which the coefficients of the series are unknown, and of truncating this series to a given number of terms. Certain coefficients may be set to zero on the basis of geometry and symmetry conditions.

The values of the unknown coefficients are then determined from a set of linear simultaneous equations which satisfy known conditions of stress, force or displacement on the boundary. The boundary point conditions may be matched exactly, or fitted in a least squares sense. Series solutions thus obtained normally satisfy some prescribed conditions in the interior of the region exactly (e.g. stress free crack conditions), and those on the other boundaries approximately.

Convergence of the collocation technique to the correct solution has not been rigorously proven. However, the technique has provided a significant number of solutions of which the accuracy may be assessed by comparison with alternative methods of solution, leading to the estimate of associated errors being less than 1% [26].

Two alternative formulations of the basic stress functions have been used for boundary collocation solution of cracked configurations, namely the Williams stress function [25], and the complex variable formulation due to Muskhelishvili [6]. Historically the Williams representation has been used for edge-cracked configurations, and the Muskhelishvili function for internally cracked configurations. Recent developments, described in Section 4.5.5, permit the extension of the collocation process based on the Muskhelishvili function, to edge-cracked configurations. With this in mind, and in order to give a proper treatment to the method, the boundary collocation procedure is described in terms of the complex stress function formulation.

4.5.1 *Collocation of the complex stress functions*

Complex variable techniques were described in Chapter 1. The stress state within a multiply connected, two-dimensional body subjected to in-plane loading may be completely specified in terms of two complex stress functions $\phi(z)$ and $\Omega(z)$. These functions, although slightly modified, derive directly from the formulation of Section 1.7.

In considering cracks which are orientated along the x-axis, the most suitable re-definition of the stress functions is:

$$\phi_*(z) = \phi'(z) \tag{4.58}$$

$$\bar{\Omega}(z) = \phi_*(z) + z\phi'_*(z) + \psi'(z) \tag{4.59}$$

see Newman [24], and Kobayashi, Cherepy and Kinsel [23]. Minor differences in notation do occur; however, the form of equations (4.58) and (4.59) has gained the most general acceptance.

The stresses and displacements are given in terms of $\phi(z)$ and $\Omega(z)$ by:

$$\sigma_x + \sigma_y = 2[\phi_*(z) + \overline{\phi_*(z)}] = 4\,\mathrm{Re}\{\phi_*(z)\} \tag{4.60}$$

$$\sigma_y - \sigma_x + 2i\tau_{xy} = 2[(\bar{z} - z)\phi'_*(z) - \phi_*(z) + \bar{\Omega}(z)] \tag{4.61}$$

$$2\mu(u + iv) = \kappa \int_0^z \phi_*(z)\,\mathrm{d}z - \int_0^{\bar{z}} \Omega(\bar{z})\,\mathrm{d}\bar{z} - (z - \bar{z})\overline{\phi_*(z)}. \tag{4.62}$$

Following the formulations of Vooren [27] and Newman [24], it can be shown that, for a region having (see Fig. 4.9):

(a) An external boundary S_o
(b) A single internal boundary S_i surrounding the origin, and subjected to a zero resultant force.
(c) A stress free crack located along the real axis with tips at distance a and b from the origin.

The complex stress functions take the form:

$$\phi_*(z) = \sum_{-\infty}^{\infty} \frac{A_m z^m}{[(z-a)(z-b)]^{1/2}} + \sum_{-\infty}^{\infty} B_m z^m \tag{4.63}$$

$$\Omega(z) = \sum_{-\infty}^{\infty} \frac{A_m z^m}{[(z-a)(z-b)]^{1/2}} - \sum_{-\infty}^{\infty} \overline{B_m z^m}. \tag{4.64}$$

DETERMINATION OF STRESS INTENSITY FACTORS

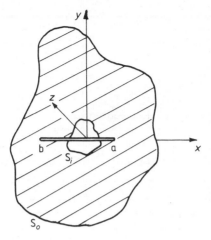

Fig. 4.9 *Two-dimensional, multiply connected, cracked body.*

In general, the coefficients A_m, B_m are complex.

The complex stress function formulation (equations (4.63) and (4.64)) imposes the equilibrium and main compatibility conditions. For an internal crack it is also necessary to satisfy a supplementary compatibility condition to ensure single-valuedness of displacements in a multiply connected domain. In a region of connectivity J this condition can be stated as:

$$\kappa \oint_{S_j} \phi(z)\,dz - \oint_{S_j} \Omega(\bar{z})\,d\bar{z} = 0, \quad j = 1, 2 \ldots, J \tag{4.65}$$

where S_j is the contour around each separate boundary. In terms of the coefficients in the series stress functions (4.63) and (4.64), the stress intensity factors in (4.1) for tip at $z = a$ are given by:

$$K_I - iK_{II} = \frac{2(2\pi)^{1/2}}{(a-b)^{1/2}} \sum_{-\infty}^{\infty} A_m a^m. \tag{4.66}$$

4.5.2 *Symmetry and anti-symmetry properties*

It was mentioned earlier that the coefficients in the series stress functions (4.63) and (4.64) may, in general, be complex. However, under uniaxial symmetry or anti-symmetry conditions these coefficients are either purely imaginary or purely real. The symmetry properties are summarized in Fig. 4.10. A knowledge of these properties should considerably reduce the computational effort involved.

66 THE MECHANICS OF FRACTURE AND FATIGUE

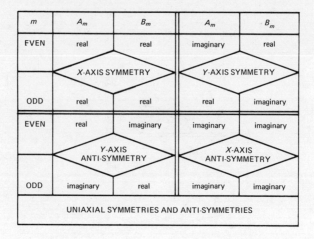

Fig. 4.10 *Symmetry properties (collocation formulation)*.

4.5.3 *Special cases of the geometry*

The position of the external boundary (whether finite or infinite), and the existence or otherwise of an internal boundary affects the form of the stress functions. Coefficients A_m, B_m in equations (4.63) and (4.64) can be set to zero for certain types of geometry. These conditions are summarized in Table 4.1. In case (1), for stress components at infinity to remain finite, then A_1, B_0, may

DETERMINATION OF STRESS INTENSITY FACTORS

Table 4.1

	Geometry description	A_m non-zero	B_m non-zero
(1)	Infinite sheet, single internal boundary	$1 \geqslant m \geqslant -\infty$	$0 \geqslant m \geqslant -\infty$
(2)	Finite sheet, no internal boundary	$1 \leqslant m \leqslant \infty$	$0 \leqslant m \leqslant \infty$
(3)	Finite sheet, single internal boundary	$\infty \geqslant m \geqslant -\infty$	$\infty \geqslant m \geqslant -\infty$

be expressed in terms of the applied stresses at infinity, namely:

$$A_1 = \sigma/2 \qquad (4.67)$$

$$B_0 = (\lambda - 1)\sigma/4 \qquad (4.68)$$

where λ is the ratio of the applied stresses at infinity.

4.5.4 Outline programming technique

Collocation points are normally selected so as to be equally distributed between internal and external boundaries (where appropriate) and to have equal angular spacing with reference to the origin of the coordinate system. This arrangement may have the advantage of producing a higher density of collocation points along vertical boundaries at points close to the crack tip. Collocation point coordinates are taken in turn, converted to complex number form, and inserted into the truncated form of the complex stress functions given in equations (4.63) and (4.64). These are then used to produce details of the relevant stresses via equations (4.60) and (4.61). Each collocation point produces one row of the main matrix $\mathcal{A}$ (see Fig. 4.11), the number of columns depending on the maximum power allowed in the truncated series.

Information on the direct and shear stresses is inserted into column vector S. One (or two) rows may be left vacant for single-valued displacement information based on equation (4.65), in cases where single-valued displacements are not automatically satisfied.

The method of filling the matrices for the qth collocation point is illustrated in Fig. 4.11. For each power in the range $-N$ to N, evaluation of (say) power r will fill the rth element of the qth row in the A_m section, and a similar element in the B_m section, and another two such elements in the (q + number of collocation points)th row. Subsequently, the qth element of the stress vector (S)

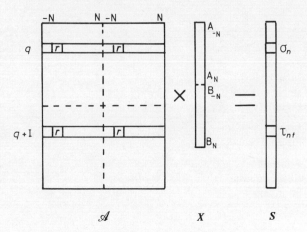

Fig. 4.11 *Form of boundary collocation matrices.*

is filled with the direct stress value, and the (q + number of collocation points)th with the shear stress value.

The resulting set of simultaneous equations may have equal numbers of equations and unknowns, requiring an exact solution. However, if additional collocation points are included, a least squares solution may be obtained. There is evidence to suggest that the latter technique is both more efficient and more accurate [24, 29]. Finally, the coefficients from the solution vector may be substituted into equation (4.66) in order to obtain the stress intensity.

The boundary collocation technique requires computer capacity of approximately 100k words. Solutions may be obtained rapidly if a packaged program is available, however the development of such a program would involve at least twelve months work for someone unfamiliar with the technique.

4.5.5 *Mapping–collocation techniques*

Considerable attention has recently been focused on the modified mapping–collocation (MMC) technique. A significant number of solutions have been obtained using this technique. The method is reviewed by Bowie [28]. The essential feature of the technique is the selection of a simple, closed-form mapping function which transforms the physical region into a singularity-free parameter region. Traction-free conditions on the mapped crack are ensured by demanding continuity of the stress function within a specified zone. A collocation process is then used in the mapped plane, with the advantage of a far simpler form of the series stress function since the singular terms are now confined to the mapping function. It is then a straightforward application of the

DETERMINATION OF STRESS INTENSITY FACTORS

inverse mapping to derive stress intensity factors from the stresses in the parameter region.

Tracy [19] reports limitations of the MMC method. Solutions for panels with a large height/width ratio present convergence problems when a single power series expansion is used to represent stress functions. Bowie and co-workers have overcome the difficulty in this case by considering such awkward configurations as being 'partitioned' into smaller panels (Fig. 4.12). Each partition is then allocated a separate power series expansion to represent the stress function in its particular area. The solutions are then 'stitched' at the interfaces so as to satisfy requirements of force and displacement compatibility over the boundaries between each 'stitched' collocation point.

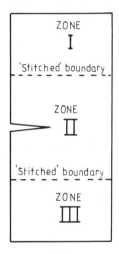

Fig. 4.12 *Partitioning of eccentric geometry (after [19])*.

By requiring continuity across a line perpendicular to the crack through its centre, Tracy [19] has introduced a stress-free boundary containing a single edge-crack. This technique has several important features:

(a) It permits collocation along the crack line, of particular importance in the modelling of residual stress effects.
(b) The edge-cracked region may be further transformed to (say) a C-shaped beam by a simple additional mapping.
(c) The solution at very short crack lengths tends to the correct, stress concentration effect, limiting value. Engineering components normally spend a large proportion of their life after crack initiation in the short crack length regime.

Thus, this mapping–collocation technique represents a powerful method for applying complex stress function methods to edge-crack problems with high accuracy at short crack lengths, whilst retaining the option of additional mappings.

4.6 Finite element methods

The finite element method is described in detail elsewhere [30]. General reviews relating to fracture mechanics appear in [26] and [31], and an extensive and up to date review of techniques applicable to fracture mechanics has been produced [32].

The basis of the finite element method is to replace the continuum geometry by an assembly of structural elements, in the case of two-dimensional elasticity these are normally triangular having three or six nodes, or quadrilateral with four or eight nodes. Each element is connected at nodal points to adjacent elements. Conditions of compatibility are satisfied at nodal points, and those of equilibrium by an energy minimization procedure. Possible stress and displacement variations are restricted within each element. The overall effect is to reduce the problem from one having an infinite number of degrees of freedom, to one having a limited number.

The solution of moderately complicated problems by the finite element method requires at least a medium capacity (100k words) electronic computer. The solution process may be divided into three basic stages:

(a) Determination of the stiffness matrix for each element. This matrix relates each nodal force (a vector) to the nodal displacement (also a vector) via the element stiffness matrix.
(b) Combination of element stiffness matrices into an overall assemblage stiffness matrix for the complete problem.
(c) Solution of the problem by using the assemblage stiffness matrix to obtain unknown forces and displacements from the known boundary conditions. Note that the resulting set of simultaneous equations which are solved normally contain a large number of zero matrix elements, and hence the overall computing requirement may be considerably reduced.

The techniques for determining stress intensity factors by finite element methods have developed rapidly over a period of a decade. With hindsight they may be classified under two main headings: non-singular crack tip representations and singular elements.

4.6.1 *Non-singular crack tip representations*

Early direct methods involved the use of a very high density of conventional elements around the crack tip [33, 34]. Stress intensity factors were derived via

DETERMINATION OF STRESS INTENSITY FACTORS

equations (3.37), for example:

$$K_I = (2\pi r)^{1/2} \lim_{r \to 0} \sigma_y \tag{4.69}$$

$$K_I = \lim_{r \to 0} \frac{Ev}{4(1-v^2)}\left(\frac{2\pi}{r}\right)^{1/2}, \quad \text{plane strain}. \tag{4.70}$$

The solution technique involves deriving the stress, σ_y, or displacement, v, at some small distance, r, from the crack tip. The latter (displacement) method is consistently more accurate than the stress method [35], and typically yields stress intensity values within 5% of accurate solutions.

In an attempt to reduce the requirement for large numbers of very small elements near the tip of the crack alternative techniques were devised which obviate the need to approach the crack tip. One method employed the expression for strain energy release rate per unit thickness:

$$G = \partial U/\partial a \tag{4.71}$$

by solving for the strain energy of the system, U, at a given crack length, a, and then by disconnecting the adjacent node to give a small increment of crack length, the change in strain energy may be calculated. The value of K may be calculated via equations (3.14) or (3.15) and typical errors are of order 2%. Alternatively the energy release rate may be calculated from the work done in closing the crack tip by a small amount [36].

Stress intensity may also be derived from the line integral

$$J = \int_\Gamma \left(U\,dy - T \cdot \frac{\partial u}{\partial x}\,ds \right) \tag{4.72}$$

which is Rice's J integral, referred to in Section 6.7.2. The J integral is a constant for any contour Γ surrounding the crack tip, U is the strain energy density, T the traction vector defined according to the outward normal along Γ, u the displacement vector and ds the arc length along Γ. For linear elasticity J is equal to G, and hence, from equations (3.14) and (3.15):

$$K_I = \begin{cases} (JE)^{1/2}, & \text{plane stress} \tag{4.73a} \\ \left(\dfrac{JE}{1-v^2}\right)^{1/2}, & \text{plane strain} \tag{4.73b} \end{cases}$$

This method has been employed by Chan, Tuba and Wilson [33], solutions being within $3\frac{1}{2}$% where comparison is possible, and of high numerical efficiency [37]. Other contour integrals have been proposed [38, 39] but will not be examined here.

The concept of compliance, C, was mentioned in Section 2.4. For an applied load P it is possible to express the strain energy release rate, per unit thickness, G, as:

$$G = \frac{P^2}{2} \frac{\partial C}{\partial a} \tag{4.74}$$

(see appendix at end of chapter for proof). Thus, substituting into equations (3.14) and (3.15):

$$K_I = \begin{cases} P\left(\dfrac{E}{2} \dfrac{\partial C}{\partial a}\right)^{1/2}, & \text{plane stress} \quad (4.75) \\[2ex] P\left(\dfrac{E}{2(1-v^2)} \dfrac{\partial C}{\partial a}\right)^{1/2}, & \text{plane strain} \quad (4.76) \end{cases}$$

In order to determine K_I from the above equation it is necessary to obtain the compliance for a range of crack lengths, and hence the derivative. The overall errors associated with this process may be estimated from [40] at $2-3\frac{1}{2}\%$.

4.6.2 Singular elements

The techniques described above depend on methods which allow the crack tip singularity to be inferred from information obtained remote from the tip. In addition they require high or medium density element meshes in the vicinity of the crack tip. An obvious extension, which would increase accuracy and reduce the necessary computational facilities, is to model the singularity at the tip of the crack. Details of the various techniques are contained in [32]. The main formulations will be considered under two headings.

Classical-solution and polynomial based functions
The near-tip stress and displacement fields of equations (3.37) to (3.42) may be written in terms of stress functions due to Westergaard [41], Muskhelishvili [6] or Williams [25]. A particular advantage of these fields is that they may be expressed directly in terms of the stress intensity factor, thus obviating the need for indirect derivation techniques.

Byskov [42] developed an element based on the Muskhelishvili function, requiring many terms in the series expression, and consequently many nodal points on the element periphery. The reader will note the similarity between this and the collocation procedure of Section 4.5. In effect, the special element is a small collocation zone.

The Williams function formed the basis of special elements used in several solutions [43, 44]. Typical errors associated with these elements are approximately 2%. The need for accurate solutions for short edge-cracks in engineering configurations has already been alluded to. In general, none of the classical-solution based functions so far employed yields the correct solution as crack

DETERMINATION OF STRESS INTENSITY FACTORS

length tends to zero. However, we have already seen in Section 4.5.5 that a mapping collocation formulation [19], which includes certain continuity requirements, does not possess this limitation. It is opined that the latter formulation may be employed in the development of a special element capable of high accuracy at very short crack lengths.

Polynomial displacement functions have also been used in order to ease the numerical solution process, appropriate singularities being introduced either by manipulation of element shapes [45] or of displacement fields [46].

Isoparametric representation

The modification of a conventional eight-noded quadrilateral element (Fig. 4.13(a)) by moving some nodes to the locations shown in Fig. 4.13(b), introduces a singularity of order $r^{-1/2}$ on all rays emanating from the corner point [47]. Further development of this technique allows additional elements adjacent to the near-tip element to include the effects of the nearby singularity [48].

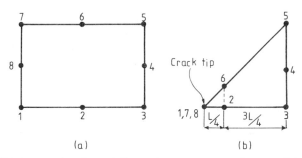

Fig. 4.13 (a) *Conventional quadrilaterial isoparametric element;* (b) *modified element with $r^{-1/2}$ singularity.*

4.7 Integral equations

Integral transforms have been used to derive stress intensity solutions for several cracked configurations [49, 50]. Mellin transform techniques [51] allow the solution of problems involving radially oriented cracks in arrays or emanating from circular holes. In addition, configurations involving radial and circumferential cracks in discs are tractable. Examples of the range of solutions are contained in [52], together with details of the method. The technique has two important advantages, namely relatively modest computational requirements (the main matrix typically contains only 40 rows and 40 columns), and high accuracy (typically less than $\frac{1}{2}$% error) even at very short crack lengths.

4.7.1 Formulation

The general elastic boundary value problem may be reduced to the solution of an integral equation of the form:

$$\int_a^b \frac{M(r, t)P(t)}{[(t-a)(b-t)]^{1/2}} \, dt = -f(r) \tag{4.77}$$

where r is an arbitrary point on the crack which lies between a and b and $f(r)$ is the crack line loading in the unflawed state. The kernel is of the form:

$$M(r, t) = \frac{1}{t-r} + F(r, t) \tag{4.78}$$

where $F(r, t)$ is not singular for $a < t < b$ and $a < r < b$. The unknown function $P(t)$ is related to v, the normal displacement of the crack surfaces by:

$$v(r) = -\frac{2(1-v^2)}{E} \int_r^b \frac{P(t)}{(t-a)(b-t)} \, dt. \tag{4.79}$$

K may be determined from the shape of the crack via equation (3.37). For tip a; the stress intensity K_a is:

$$K_a = \frac{E}{2(1-v^2)} \lim_{r \to a^+} \left\{ [2\pi(r-a)]^{1/2} \frac{\partial v(r)}{\partial r} \right\} \tag{4.80}$$

Finally, comparing equations (4.79) and (4.80) we note that

$$K_a = P(a).$$

Similarly:

$$K_b = P(b)$$

hence, a solution of equation (4.77) for P as a function of crack position leads directly to the required stress intensity. The integral equation (4.77) may be solved in some special cases by conventional Gauss–Chebyshev quadrature [52], or by a modified form [52, 53].

4.8 Boundary methods

Boundary methods are variously described as boundary element methods, boundary integral equation methods or body force methods. They are one of the current 'growth' areas in the numerical modelling of engineering problems. The techniques will be described in general terms, since the necessary formulation, deriving from Navier's equation for three-dimensional elasticity is not suited to an introductory text.

In general, numerical solution techniques fall into two main categories: differential methods, including finite element (FE) methods, and integral

DETERMINATION OF STRESS INTENSITY FACTORS

methods. Since numerical integration is inherently more accurate than differentiation, provided the problems of formulation and programming can be overcome, there may be advantages to integral methods. In fact, several distinct advantages are apparent [54]:

(a) Because only the boundaries of the body need be discretized, very much smaller systems of algebraic equations are generated than with the FE method, with its internal nodes. (Whilst the matrices generated are much smaller, they are fully populated. However, provided the ratio of boundary length/area (2-D), boundary area/volume (3-D) is not excessive, boundary methods retain their superiority.)
(b) Because there is no internal subdivision, no approximation is imposed on the solution at interior points, leading to higher stress accuracy than with FE.
(c) Values of solution variables are obtained at a limited number of points, and it is possible to concentrate on particular regions of interest (stress concentrations, cracks, interfaces, etc.). It is not necessary to obtain solutions at a large number of internal locations which are of limited interest.
(d) The basic formulation is identical for two and three dimensions, so that the method is particularly effective for three-dimensional regions.
(e) Conditions on extreme boundaries, located at infinity, may be satisfied automatically.

The technique involves superposition of particular solutions of the governing differential equations. The basic analytical solutions used are those for a unit excitation (e.g. force) applied to the larger region in which the physical region is 'embedded'. Boundary element methods have been reviewed elsewhere [54, 55]. In the direct formulation the unknown functions in the integral equation are the physical stresses and/or displacements on the boundaries. Solution of the integral equation gives stresses on the boundary directly, and those elsewhere are obtained by numerical integration. Workers commonly associated with this approach include Cruse [56] and Lachat [57].

In the case of indirect methods, fictitious singular sources are distributed along appropriate boundaries, and the problem is solved in terms of the source densities, such that specified boundary conditions are satisfied around the physical, embedded region [58]. This approach is similar to the so-called body force method [59, 60]. The formal equivalence of direct and indirect methods has been considered by Brebbia and Butterfield [61]. Finally, for the semi-direct method, the integral equations are formed in terms of unknown functions related to stress functions [62].

Recent crack problem solutions using boundary methods have involved the extrapolation of crack face displacements via equation (3.37), as in Section 4.6.1, in order to obtain stress intensity. The errors associated with this method are estimated tentatively at 5%. The indirect method, combined with conformal mapping techniques, was applied to a plane cracked problem [63] with encouraging results (maximum error approximately $1\frac{1}{2}\%$).

Thinking in terms of the indirect method, and the solution for point force density around the embedded physical boundary, it is clear that by employing analytic solutions for two- and three-dimensional regions containing cracks, it may be possible to produce solutions of high accuracy, since the boundaries on which the distributions are approximated do not include the crack surfaces.

4.9 The compounding method

The compounding method is a particularly rapid, approximate method for extending the range of available solutions. In addition it is possible to put error bounds on compounded results with some confidence. The derivation of solutions using this method will involve time expenditure of a few hours.

A cracked configuration may have several boundaries, e.g. sheet edges (of infinite radius), holes (of finite radius) and other cracks (of zero radius), all of which will exercise their own effect on the stress intensity factor.

Considering the configuration to be separated into a number of ancillary configurations, each having the same general loading, and a known solution, Cartwright and Rooke [64] have shown that these separate solutions may be 'compounded' to produce the required solution, K_r, in accordance with the equation:

$$K_r = K_0 + \left(\sum_{n=1}^{N} (K_n - K_0) \right) + K_e \qquad (4.81)$$

where K_0 is the stress intensity factor in the absence of all boundaries, of a form applicable to the loading. Thus $K_0 = \sigma(\pi a)^{1/2}$ for remote loading or a pressurized crack, and $K_0 = P/(\pi a)^{1/2}$ for point loading. K_n is the stress intensity factor for the nth ancillary configuration, and K_e accounts for the effects of interaction between boundaries.

Solutions are frequently presented in terms of Q, the configuration correction factor, where:

$$Q_n = K_n/K_0. \qquad (4.82)$$

Therefore, it is convenient to express (4.81) in the non-dimensional form:

$$\frac{K_r}{K_0} = 1 + \left[\sum_{n=1}^{N} \left(\frac{K_n}{K_0} - 1 \right) + \frac{K_e}{K_0} \right]. \qquad (4.83)$$

The only unknown on the right-hand side of (4.83) is K_e. For most configurations in which the ancillary boundaries are well separated from each other, relative to the crack size, neglecting the term K_e/K_0 will lead to underestimates in the

DETERMINATION OF STRESS INTENSITY FACTORS

compounded solution, K_r/K_0, of less than 10%. These underestimates may be significantly less than 10% in certain circumstances. In important engineering problem of cracks emanating from hole boundaries, and configurations involving large hole diameter/sheet width ratios the K_e/K_0 term may be estimated [65, 66]. Furthermore, solutions for cracked, stiffened structures may also be compounded with underestimates of less than 5% [67].

The compounding technique, together with various worked examples for the cases in which K_e/K_0 is negligible, is described in a recent data item [68]. The accuracies associated with the basic compounding method are assessed for three configurations which include the whole range of possible radii in ancillary configurations (Fig. 4.14). By comparison with available solutions, the percentage underestimates associated with each compounded solution are shown in Table 4.2.

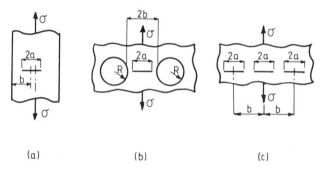

Fig. 4.14 *Examples of compounding:* (a) *straight boundaries;* (b) *circular boundaries;* (c) *other crack boundaries.*

Table 4.2

a/b	Straight boundaries (%)	Circular boundaries (%)	Crack boundaries (%)
0.2	2	2	1
0.4	4	2	2
0.6	9	3	–
0.8	25	6	–

Example 4.4

A complex configuration develops a crack, as illustrated in Fig. 4.15(a). Employ the compounding method to estimate the stress intensity at crack tips A and B.

78 THE MECHANICS OF FRACTURE AND FATIGUE

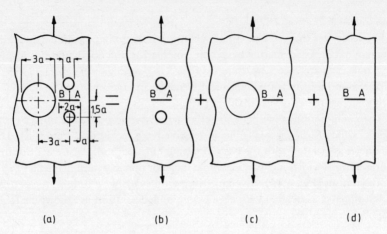

Fig. 4.15 *Compounding method:* (a) *problem configuration;* (b)–(d) *ancillary configurations (Example 4.4).*

Solution: Divide the problem into ancillary configurations (Fig. 4.15(b) to (d)). Solutions for these ancillary configurations, are contained in a standard reference work [69], Sections 1.3.7, 1.3.5, and 1.1.11 respectively, and are rapidly extracted as:

$$\left.\frac{K_b}{K_0}\right|_{\text{tip A}} = 0.880 \qquad \left.\frac{K_b}{K_0}\right|_{\text{tip B}} = 0.880$$

$$\left.\frac{K_c}{K_0}\right|_{\text{tip A}} = 1.201 \qquad \left.\frac{K_c}{K_0}\right|_{\text{tip B}} = 1.561$$

$$\left.\frac{K_d}{K_0}\right|_{\text{tip A}} = 1.091 \qquad \left.\frac{K_d}{K_0}\right|_{\text{tip B}} = 1.056$$

which may be substituted into equation (4.83), neglecting K_e/K_0, to give:

$$\left.\frac{K_r}{K_0}\right|_{\text{tip A}} = 1 + (-0.12 + 0.201 + 0.091) = 1.172$$

$$\left.\frac{K_r}{K_0}\right|_{\text{tip B}} = 1 + (-0.12 + 0.561 + 0.056) = 1.497.$$

4.10 Experimental methods

It is not intended to consider experimental methods in depth, except insofar as they amplify the stress analysis or design techniques and their limitations.

DETERMINATION OF STRESS INTENSITY FACTORS

Experimental methods have been reviewed elsewhere [31].

Briefly, stress intensity may be derived from an experimentally measured change in *compliance* with crack length via equations (4.75) or (4.76). In the case of *photoelasticity* and *interferometry*, limiting stress values in the near-tip region are associated with stress intensity. Finally, *fatigue crack growth* results from complex configurations (see Chapter 7) may be employed in stress intensity derivation via crack growth rate data obtained from specimens with known stress intensity calibrations.

Appendix A: Uniqueness of the weight function

Consider a cracked body having a crack of length a, and acted upon by loads P_1, $P_2, P_3, \ldots, P_N$. Confining our attention to loads P_1 and P_2, and recalling the definition of compliance (Section 2.4), the displacement of loading points is given by:

$$u_1 = C_{11}P_1 + C_{12}P_2 = u_1^1 + u_1^2 \tag{4.84}$$

$$u_2 = C_{22}P_2 + C_{21}P_1 = u_2^2 + u_2^1 \tag{4.85}$$

where u_i is the displacement of P_i, u_i^j is the component of u_i caused by P_j. The term C_{ij} is the displacement coefficient (a function of crack length) which gives the displacement at i caused by a unit load at j. The reciprocal theorem states that:

$$C_{ij} = C_{ji}. \tag{4.86}$$

Knowing forces and displacements, we calculate the strain energy, U, resulting from the movement of each load

$$U|_{P_1} = \tfrac{1}{2} P_1 U_1 = \tfrac{1}{2} P_1 (C_{11}P_1 + C_{12}P_2) \tag{4.87}$$

$$U|_{P_2} = \tfrac{1}{2} P_2 U_2 = \tfrac{1}{2} P_2 (C_{22}P_2 + C_{21}P_1). \tag{4.88}$$

Summing the separate strain energies the total strain energy is:

$$U = \tfrac{1}{2}(C_{11}P_1 P_1 + C_{12}P_1 P_2 + C_{22}P_2 P_2 + C_{21}P_1 P_2) \tag{4.89}$$

Noting the results of Section 2.4.1:

$$G = \left.\frac{\partial U}{\partial a}\right|_{\text{fixed loads}}. \tag{4.90}$$

Thus, substituting from (4.89) into (4.90)

$$G = \tfrac{1}{2}\left(\frac{\partial C_{11}}{\partial a} P_1 P_1 + \frac{\partial C_{12}}{\partial a} P_1 P_2 + \frac{\partial C_{22}}{\partial a} P_2 P_2 + \frac{\partial C_{21}}{\partial a} P_1 P_2\right). \tag{4.91}$$

The total stress intensity, K, may be obtained by the superposition of K_1 and

K_2, the stress intensities due to loads P_1 and P_2 respectively. Therefore:

$$K = K_1 + K_2. \tag{4.92}$$

(Caution: the suffixes do not refer to the mode of crack deformation, but to separate components of the same mode.) Now define:

$$k_i = K_i/P_i \tag{4.93}$$

by substitution into (4.91) we obtain:

$$K = k_1 P_1 + k_2 P_2. \tag{4.94}$$

Furthermore, from equation (4.8):

$$G = K^2/H \tag{4.95}$$

which yields, on substitution from (4.92):

$$G = (k_1 k_2 P_1 P_2 + k_1 k_1 P_1 P_1 + k_2 k_1 P_2 P_1 + k_2 k_2 P_2 P_2) \tag{4.96}$$

equating like coefficients in (4.91) and (4.96):

$$\begin{aligned} \frac{k_1 k_2}{H} &= \tfrac{1}{2}\frac{\partial C_{12}}{\partial a}, & \frac{k_2 k_1}{H} &= \tfrac{1}{2}\frac{\partial C_{21}}{\partial a} \\ \frac{k_1 k_1}{H} &= \tfrac{1}{2}\frac{\partial C_{11}}{\partial a}, & \frac{k_2 k_2}{H} &= \tfrac{1}{2}\frac{\partial C_{22}}{\partial a}. \end{aligned} \tag{4.97}$$

Re-arranging two of these results

$$k_1 = \frac{H}{2}\frac{\partial C_{12}}{\partial a}\frac{1}{k_2}, \quad k_2 = \frac{H}{2}\frac{\partial C_{22}}{\partial a}\frac{1}{k_2}. \tag{4.98}$$

Substituting from (4.93):

$$k_1 = \frac{H}{2}\frac{\partial C_{12}}{\partial a}\frac{P_2}{K_2}, \quad k_2 = \frac{H}{2}\frac{\partial C_{22}}{\partial a}\frac{P_2}{K_2}. \tag{4.99}$$

Combining equations (4.92), (4.93) and (4.99):

$$K = \frac{H}{2}\frac{P_2}{K_2}\left(\frac{\partial C_{12}}{\partial a}P_1 + \frac{\partial C_{22}}{\partial a}P_2\right) \tag{4.100}$$

which may be generalized, for the case of N applied loads, to [70]:

$$K = \frac{H}{2}\frac{P_m}{K_m}\sum_{i=1}^{N}\frac{\partial C_{im}}{\partial a}P_i. \tag{4.101}$$

Substituting from (4.84) and (4.85) into (4.100):

$$K = \frac{H}{2}\frac{1}{K_2}\left(\frac{\partial u_1^2}{\partial a}P_1 + \frac{\partial u_2^2}{\partial a}P_2\right). \tag{4.102}$$

DETERMINATION OF STRESS INTENSITY FACTORS

The implications of (4.100), (4.101) and (4.102) are somewhat surprising. Stress intensity results can be found for any loading from the results for one load. For arbitrary distributed tractions t over surface Γ the generalized form of the results becomes:

$$K = \int_\Gamma t \cdot h \, d\Gamma \qquad (4.103)$$

where

$$h = h(x, y, a) = \frac{H}{2K_m} \frac{\partial u^m}{\partial a}.$$

Appendix B: The weight function for mixed boundary condition problems

Consider the cracked configuration illustrated in Fig. 4.16, having displacement boundary conditions over S_U and stress boundary conditions over the remainder of the boundary, S_T. Bowie and Freese [71] have modified the formulation due to Rice [10] to allow the applied tractions to be functions of both position and crack length, demonstrating that:

$$K^{(1)}K^{(2)} = \frac{H}{2} \int_{S_T + S_U} \left[T^{(2)} \frac{\partial U^{(1)}}{\partial a} - U^{(2)} \frac{\partial T^{(1)}}{\partial a} \right] dS \qquad (4.104)$$

where $K^{(i)}$, $(i = 1, 2)$ is the crack tip stress intensity factor for a loading system (i) which consists of $T^{(i)}$ applied over S_T, and displacements $U^{(i)}$ applied over S_U. In general $K^{(i)} = K^{(i)}(a)$, $T^{(i)} = T^{(i)}(x, y)$ and $U^{(i)} = U^{(i)}(x, y)$. In order to investigate the implications of (4.104) we look at two particular loading systems.

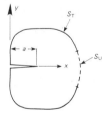

Fig. 4.16 *Cracked configuration used in formulation.*

First loading system

In the event that $\partial T^{(1)}/\partial a = 0$ over S_T (representing stress boundary conditions alone) and $U^{(2)} = 0$ over S_U (fixed-grip conditions) equation (4.104) gives us:

$$K^{(2)} = \int_{S_T} T^{(2)} m(x, y, a) \, dS \qquad (4.105)$$

where:

$$m(x, y, a) = \frac{H}{2K^{(1)}} \frac{\partial U^{(1)}}{\partial a} \qquad (4.106)$$

and $m(x, y, a)$ is the familiar weight function of Bueckner [9] and Rice [10].

Second loading system

In this case $\partial U^{(1)}/\partial a = 0$ over S_U (representing displacement boundary conditions) and $T^{(2)} = 0$ over S_T (traction-free boundary). From equation (4.104) we obtain:

$$K^{(2)} = \int_{S_U} U^{(2)} m^*(x, y, a) \, dS \qquad (4.107)$$

where:

$$m^*(x, y, a) = \frac{-H}{2K^{(1)}} \frac{\partial T^{(1)}}{\partial a} \qquad (4.108)$$

The implications of equations (4.105–4.108) are that it is possible to obtain a conventional weight function, $m(x, y, a)$ with $U = 0$ along S_U and arbitrary tractions applied to S_T. Furthermore, one may obtain an additional function, $m^*(x, y, a)$ with $T = 0$ along S_T, for arbitrary displacements along S_U.

In order to solve the mixed boundary condition problem, loading system 'A' in Fig. 4.17, consisting of tractions $\tilde{T}(x, y)$ along S_T and displacements $\bar{U}(x, y)$ along S_U, we invoke the linear superposition illustrated in Fig. 4.17(b) and (c), wherein system 'B' has displacements set to zero over S_U and system 'C' has tractions set to zero over S_T. Denoting stress intensity factors for the three systems as K_A, K_B and K_C respectively, we see:

$$K_A = K_B + K_C \qquad (4.109)$$

clearly, K_B may be obtained from equation (4.105) as:

$$K_B = \int_{S_T} \tilde{T}(x, y) \, m(x, y, a) \, dS \qquad (4.110)$$

DETERMINATION OF STRESS INTENSITY FACTORS

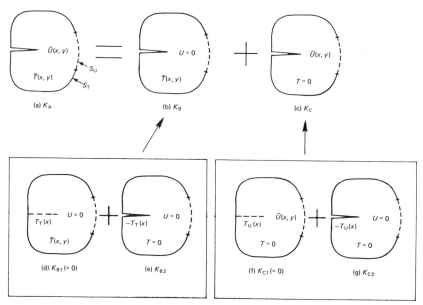

Fig. 4.17 *Superpositions for mixed boundary condition problems.*

whilst K_C is given by equation (4.107) as:

$$K_C = \int_{S_U} \bar{U}(x, y) \, m^*(x, y, a) \, dS \qquad (4.111)$$

This approach requires a knowledge of m for the evaluation of K_B and of m^* for the evaluation of K_C. As an alternative, system 'C' may be solved by the additional superposition $K_{C1} + K_{C2}$, illustrated in Fig. 4.17, which requires a knowledge of m and of the crack-line loading in the unflawed structure, $T_U(x)$ caused by displacement boundary conditions alone.

Finally, an alternative approach to the solution of system 'B' is the super superposition $K_{B1} + K_{B2}$, which again requires a knowledge of m and of the crack-line loading in the uncracked structure, $T_T(x)$ arising from traction boundary conditions alone. In the event that the crack-line loading due to non-displacement boundary conditions includes a contribution from residual stresses or body forces, the latter superposition may be the most straightforward solution method.

Methods of determining m and m^*, and various possible applications, are discussed in [72].

References

1. Irwin, G. R. (1958), *Fracture Handbuch der Physik, VI*, 551–590, Springer-Verlag, Heidelberg.
2. Irwin, G. R. (1960), *Structural Mechanics*, pp. 557–94, Pergamon, New York.
3. Tada, H. (1970), 'Westergaard stress functions for several periodic crack problems', *Engng Frac. Mech.*, 2, 177–80.
4. Sih, G. C. (1966), 'On the Westergaard method of crack analysis', *Int. J. Frac. Mech.*, 2, 628–31.
5. Eftis, J., Subramonian N. and Liebowitz H. (1977), 'Crack border stress and displacement equations revisited', *Engng Frac. Mech.*, 9, 189–210.
6. Muskhelishvili, N. I. (1953), *Some Basic Problems of the Mathematical theory of elasticity*, Noordhoff, Leiden.
7. Paris, P. C. and Sih, G. C. (1965), 'Stress analysis of cracks' in *Fracture Toughness Testing and its Applications*, ASTM STP 381, 30–77.
8. Cartwright, D. J. and Rooke, D. P. (1979), 'Green's functions in fracture mechanics' in *Proc. Symp. on Fracture Mechanics – Current Status, Future Prospects, University of Cambridge*, March.
9. Bueckner, H. F. (1970), 'A novel principle for the computation of stress intensity factors', *Z. Angewandte Mathemat. Mechan.*, 50, No. 9, 529–46.
10. Rice, J. R. (1972). 'Some remarks on elastic crack-tip stress fields', *Int. J. Solids Structures*, 8, No. 6, 751–8.
11. Sneddon, I. N. and Srivastav, R. P. (1963–4), 'The stress in the vicinity of an infinite row of collinear cracks', *Proc. R. Soc. Edinburgh* A, 67, Pt. I, 39–41.
12. Parker, A. P. (1979), *Mechanics of Fracture and Fatigue in Some Common Structural Configurations*, RMCS Technical Note, MAT/18.
13. Bueckner, H. F. (1971), 'Weight functions for the notched bar', *Z. Angewandte Mathemat. Mechan.*, 51, 97–109.
14. Tada, H., Paris, P. and Irwin, G. (1973), *The Stress Analysis of Cracks Handbook*, Del Research Corp., Hellertown, Pennsylvania.
15. Bowie, O. L. and Freese, C. E. (1976), 'On the 'overlapping' problem in crack analysis', *Engng Frac. Mech.*, 8, 373–9.
16. Grandt, A. F. (1975), *Two dimensional stress intensity factor solutions for radially cracked rings*, AFML-TR-75-121, Wright-Patterson AFB, Ohio.
17. Impellizzeri, L. F. and Rich, D. L. (1976), *Spectrum fatigue crack growth in lugs*, ASTM STP 595, 320–36.
18. Grandt, A. F. (1978), 'Stress intensity factors for cracked holes and rings loaded with polynomial crack face pressure distributions', *Int. J. Frac.*, 14, R221–9.
19. Tracy, P. G. (1975), 'Analysis of a radial crack in a circular ring segment', *Engng Frac. Mech.*, 7, 253–60.

20. Labbens, R. C., Heliot, J. and Pellissier-Tanon, A. (1976), *Weight Functions for Three-Dimensional Symmetrical Crack Problems*, ASTM STP 601, 448–70.
21. Petroski, H. J. and Achenbach, J. D. (1978), 'Computation of the weight function from a stress intensity factor', *Engng Frac. Mech.*, **10**, 257–66.
22. Hooke, C. J. (1970), 'Numerical solution of axisymmetric stress problems by point matching', *J. Strain Analysis*, **5**, No. 1, 25–37.
23. Kobayashi, A. S., Cherepy, R. D. and Kinsel, W. C. (1964), 'A numerical procedure for estimating the stress intensity for a crack in a finite plate', *Trans. ASME, J. Bas. Engng*, **86**, 681–4.
24. Newman, J. C. (1971), *An Improved Method of Collocation for the Stress Analysis of Cracked Plates with Various Shaped Boundaries*, NASA TN D 6376.
25. Williams, M. L. (1957), 'On the stress distribution at the base of a stationary crack', *Trans. ASME, J. Appl. Mech.*, **24**, No. 1, 109–14.
26. Cartwright, D. J. and Rooke, D. P. (1973), *Methods of Determining Stress Intensity Factors*, RAE TR 73031, RAE Farnborough.
27. Vooren, J. (1967), 'Remarks on an existing numerical method to estimate the stress intensity factor of a straight crack in a finite plate', *Trans. ASME, J. Bas. Engng*, **89**, 236.
28. Bowie, O. L. (1973), 'Solutions of plane crack problems by mapping techniques' in *Methods of Analysis and Solutions of Crack Problems*, Ed. G. C. Sih. Noordhoff Int., Leiden.
29. Eason, E. D. (1976), 'A review of least squares methods for solving partial differential equations', *Int. J. Num. Meth. Engng*, **10**, 1021–46.
30. Zienkiewicz, O. C. (1971), *The Finite Element Method in Engineering Science*, McGraw-Hill, New York.
31. Cartwright, D. J. and Rooke, D. P. (1975), 'Evaluation of stress intensity factors', *J. Strain Analysis*, **10**, No. 4, 217–24.
32. Gallagher, R. H. (1978), 'A review of finite element techniques in fracture mechanics', *Proc. 1st Int. Conf. on Numerical Methods in Fracture Mechanics Swansea, UK,* Eds A. R. Luxmoore and D. R. J. Owen.
33. Chan, S. K., Tuba, I. S. and Wilson, W. K. (1970), 'On the finite element method in linear fracture mechanics', *Engng Frac. Mech.*, **2**, 1–17.
34. Kobayashi, A. S., Maiden, D. E., Simon, J. B. and Iida S. (1969), 'Application of the mechanics of finite element analysis to two dimensional problems in fracture mechanics', Paper 69 WA-PVP-12, *ASME Winter Annual Mtg*, November.
35. Yamomoto, Y., Tokuda, N. and Sumi, Y. (1973), 'Finite element treatment of singularities of boundary value problems and its application to analysis of stress intensity factors' in *Theory and Practice in Finite Element Structural Analysis*, University of Tokyo Press, Tokyo.

36. Jerram, K. (1970), *Report* RD/B/N1521, CEGB, Berkeley Nuclear Labs.
37. Prij, J. (1977) 'Two and three dimensional finite element analysis of a central-cracked . . . ', *Trans. 4th SMIRT Conf.*, Vol. 6, Paper G4/2.
38. Stern, M., Becker, E. and Dunham, R. S. (1976), 'A contour integral computation of mixed mode stress intensity factors, *Int. J. Frac.*, **12**, 359–68.
39. Stern, M. and Soni, M. (1975), 'The calculation of stress intensity factors in anisotropic materials by a contour integral method' in *Computational Fracture Mechanics*, Eds E. Rybicki and S. Benzley., ASME Special Publication.
40. Mowbray, D. F. (1970), 'A note on the finite element method in linear fracture mechanics', *Engng Frac. Mech.*, **2**, 173–6.
41. Westergaard, H. M. (1937), 'Bearing pressures and cracks', *J. Appl. Mech.*, **6**, A49–53.
42. Byskov, E. (1970), 'The calculation of stress intensity factors using the finite element method with cracked elements', *Int. J. Frac. Mech.*, **6**, 159–67.
43. Jones, R. and Callinan, R. J. (1977), 'On the use of special crack tip elements in cracked elastic sheets', *Int. J. Frac.*, **13**, No. 1, 51–64.
44. Wilson, W. K. (1971), *Some Crack Tip Finite Elements for Plate Elasticity*, ASTM STP 514.
45. Levy, N., Marcal, P. V., Ostergren, W. and Rice, J. (1971), 'Small scale yielding near a crack in plane strain: a finite element analysis', *Int. J. Frac. Mech.*, **7**, No. 2, 143–57.
46. Tracey, D. and Cook, T. S. (1977), 'Analysis of power type singularities using finite elements', *Int. J. Num. Meth. Engng*, **11**, No. 8, 1225–35.
47. Barsoum, R. S. (1977), 'Triangular quarter point elements as elastic and perfectly plastic crack tip elements', *Int. J. Num. Meth. Engng*, **11**, No. 1, 85–98.
48. Lynn, P. and Ingraffea, A. R. (1978), 'Transition elements to be used with quarter point crack tip elements', *Int. J. Num. Meth. Engng*, **12**, 1031–6.
49. Sneddon, I. N. and Lowengrub, M. (1969), *Crack Problems in the Classical Theory of Elasticity*, Wiley, New York.
50. Sneddon, I. N. (1973), 'Integral transform methods' in *Methods of Analysis and Solutions of Crack Problems*, Ed. G. C. Sih, Noordhoff Int., Leiden.
51. Tweed, J. (1973), 'The solution of certain triple integral equations involving inverse mellin transforms', *Glasgow Math. J.*, **14**, No. 1, 65–72.
52. Rooke, D. P. (1978), 'The solution of integral equations in the determination of stress intensity factors' in *Proc. 1st Int. Conf. on Numerical Methods in Fracture Mechanics, Swansea, UK* Eds A. R. Luxmoore and D. R. J. Owen.
53. Erdogan, F. and Gupta, G. D. (1972), 'On the numerical solution of singular integral equations', *Q. J. Appl. Math*, **29**, 525–34.
54. Banerjee, P. K. and Butterfield, R. (1977), Boundary element methods in

geomechanics' in *Finite Elements in Geomechanics*, Ed. G. Gudehus, Wiley, New York.
55. Brebbia, C. A. (1978), *The Boundary Element Method for Engineers*, Pentech Press, Plymouth.
56. Cruse, T. A. (1974), 'An improved boundary integral equation method for three-dimensional elastic stress analysis', *J. Comput. & Structures*, 4, 741–57.
57. Lachat, J. C. (1975), 'Further development of the boundary integral equation methods for elasto statics', *Ph.D. Thesis*, University of Southampton.
58. Banerjee, P. K. (1976), 'Analysis of vertical pile groups embedded in non-homogeneous soil' in *Proc. 6th Eur. Conf. on Soil Mech. Found. Engng, Vienna, Austria.*
59. Nisitani, H. (1978), 'Stress concentration for the tension of a strip with edge cracks or elliptic-arc notches' in *Proc. 1st Int. Conf. on Numerical Methods in Fracture Mechanics, Swansea, UK,* Eds A. R. Luxmoore and D. R. J. Owen, pp. 67–80.
60. Isida, M. (1978), 'A new procedure of the body force method with applications to fracture mechanics' in *Proc. 1st Int. Conf. on Numerical Methods in Fracture Mechanics, Swansea, UK,* Eds A. R. Luxmoore and D. R. J. Owen, pp. 81–94.
61. Brebbia, C. A. and Butterfield, R. (1978), 'Formal equivalence of direct and indirect boundary element methods', *Appl. Math. Modelling*, 2, 132–4.
62. Rim, K. and Henry, A. S. (1969), *Improvement of an Integral Equation Method in Plane Elasticity Through Modification of Source Density Representation,* NASA CR-1273.
63. Mir-Mohammad-Sadegh, A. and Alteiro, N. J. (1979), 'Solution of the problem of a crack in a finite plane region using an indirect boundary integral method, *Engng Frac. Mech.*, 11, 831–7.
64. Cartwright, D. J. and Rooke, D. P. (1974), 'Approximate stress intensity factors compounded from known solutions', *Engng Frac. Mech.*, 6, 563–71.
65. Rooke, D. P. (1976), *Stress Intensity Factors for Cracks at the Edges of Holes*, RAE TR 76087, RAE Farnborough.
66. Rooke, D. P. (1977), 'Stress Intensity factors for cracked holes in the presence of other boundaries' in *Fracture Mechanics in Engineering Practice.* Applied Science Publishers, Barking, 149–63.
67. Rooke, D. P. and Cartwright, D. J. (1976), 'The compounding method applied to cracks in stiffened sheets' *Engng Frac. Mech.*, 8, 567–73.
68. *The Compounding Method of Estimating Stress Intensity Factors for Cracks in Complex Configurations Using Solutions for Simple Configurations* (1978), Engineering Sciences Data Unit ITEM Number 78036, November.
69. Rooke, D. P. and Cartwright, D. J. (1976), *Compendium of Stress Intensity Factors*, HMSO, London.

70. Paris, P. C., McMeeking, R. M. and Tada, H. (1976), 'The weight function method for determining stress intensity factors' in *Cracks and Fracture*, ASTM STP 601, 471–489.
71. Bowie, O. L. and Freese, C. E. (1981), 'Cracked rectangular sheet with linearly varying end displacements' *Engng Frac. Mech.* (in press).
72. Parker, A. P. and Bowie, O. L. (1981), 'The weight function for mixed boundary condition problems' (in preparation).

5 Mixed-Mode Fracture Mechanics

5.1 Introduction

The bulk of fracture mechanics work to date has been concerned with single-mode loading. Many practical situations are mixed mode, but pure mode I loading may be assumed in order to obtain solutions, which can lead to unsafe design. Conversely, the lack of appreciation of mixed-mode effects may lead to overconservative design criteria in an attempt to compensate for lack of data.

It is possible to list various mixed-mode situations (e.g. cracked bars in torsion and bending, cracks loaded by a resultant force, many welds, pressure vessel nozzles, cracked rotating turbine blades, angled cracks in pressure cabins etc.). Johnson[1] in an Apollo experience report has described several mixed-mode failures. However, geometries which start in one mode may in fact become mixed mode during the life of the structure, [2–4]. Indeed, the striking fact about the current state of fracture mechanics is how little is known about mixed-mode phenomena [5].

5.2 The effective stress intensity factor in mixed mode

It would be of value to know whether a mixed-mode situation produces an effective stress intensity factor (designated K_e) and what form this stress intensity factor takes. Such information might usefully be employed in tests for a critical effective stress intensity factor (designated K_{ec}). Many workers have, in effect, been seeking the function f in the relationship

$$f(K_I, K_{II}, K_{III}) = K_e.$$

This problem was investigated by Erdogan and Sih [6], who studied crack extension in plates under combined K_I and K_{II}, both theoretically and experimentally. Energy arguments were used to show that a fracture criterion in terms of stress intensity factors can be written

$$\frac{1}{\pi}(a_{11}K_I^2 + 2a_{12}K_I K_{II} + a_{22}K_{II}^2) = S \tag{5.1}$$

where the constant S is the strain energy density factor and the constants a_{ij} ($i, j = 1, 2$) are material dependent. The experimental investigation did, indeed, produce an ellipse-like distribution of the general form predicted by equation (5.1), (see Fig. 5.1). Wilson [7] conducted experiments involving combinations of (K_I, K_{II}) and (K_I, K_{III}). In both cases he reports similar results, which conform to the general ellipse-type distribution reported by Erdogan and Sih [6]. Tanaka [8] also used a form of equation (5.1) to define the threshold condition for cyclic crack propagation, namely:

$$A_{11}(\Delta K_I)^2 + 2A_{12}(\Delta K_I \Delta K_{II}) + A_{22}(\Delta K_{III})^2 = 1. \tag{5.2}$$

The prefix Δ refers to the range of the stress intensity factor (see Chapter 7).

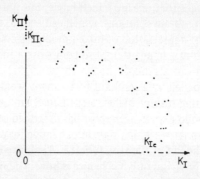

Fig. 5.1 K_I versus K_{II} at initiation of crack extension (after [6]).

Broek [5] uses the energy balance criterion (with the additional implicit assumption of in-plane crack extension) to show that the material constants in equation (5.1) reduce to

$$a_{12} = 0 \quad a_{11} = a_{22}$$

which may be substituted in (5.1) to predict that $K_{Ic} = K_{IIc}$, making the locus for combined mode cracking a circle of radius K_{Ic}. In practice $K_{Ic} \neq K_{IIc}$, and Broek suggests that the fracture condition is more likely to be of a form first proposed by Wu [9], namely:

$$\left(\frac{K_I}{K_{Ic}}\right)^2 + \left(\frac{K_{II}}{K_{IIc}}\right)^2 = 1. \tag{5.3}$$

This form fits reasonably well mixed-mode plexiglass experiments [6], DTD 5050 aluminium alloy plane strain tests [2], and 2024-T3 aluminium alloy plane stress tests [10].

Available data suggest that

$$K_{IIc} \simeq 0.75 \, K_{Ic}. \tag{5.4}$$

On this basis, we may combine equations (5.3) and (5.4) to give, as an engineering rule of thumb, the fracture criterion [5]:

$$K_I^2 + 1.78 K_{II}^2 = K_{Ic}^2. \tag{5.5}$$

5.3 Crack direction in mixed mode

In the classical Irwin theory [11] it is assumed that the crack extends in a plane coincident with that of the original crack which is contrary to experimental evidence for all except pure mode I loading. The established relationships between G_I and K_I (equations (3.14) and (3.15)):

$$G_I = \begin{cases} \dfrac{K_I^2}{E}, & \text{plane stress} \quad (5.6) \\ \dfrac{K_I^2(1-v^2)}{E}, & \text{plane strain} \quad (5.7) \end{cases}$$

are in fact the only physically acceptable combination of G_N and K_N (N = I, II, III).

Until recently, few workers have concerned themselves with the direction of crack extension or crack growth under mixed-mode conditions. It was either considered to be unimportant, or crack extension in the plane of the initial crack was tacitly assumed. However, there is already sufficient evidence to show that the direction of crack growth is neither unimportant, nor is it in plane [2, 6, 12].

Two factors which are of particular importance in considering crack direction are the type of loading (cyclic or monotonic), and the magnitude of the load in relation to the threshold load to cause fracture.

A number of workers have investigated the geometry shown in Fig. 5.2, where β represents the initial crack angle, and θ the fracture angle relative to the initial crack.

Erdogan and Sih [6] investigated angled crack extension under monotonic loading. The experiments were performed on plexiglass sheets, and the results are shown in Fig. 5.3. It was postulated that the crack was extending in a direction perpendicular to the maximum tangential stress at the crack tip, and this theoretical curve is also shown in the figure. Later, Williams and Ewing [12] performed additional work on plexiglass. In particular, the region $0° < \beta < 40°$ where various theories differ, was fully investigated. Most significantly, Williams and Ewing report at $\beta = 0°$ the points tend to $\theta = -90°$ and not $-70.5°$ as predicted by Erdogan and Sih. In their theoretical analysis Williams and Ewing [12] modify the maximum tangential stress theory of Erdogan and Sih [6] by taking account of terms in the series expression in addition to the singular term, as discussed by Cotterell [13].

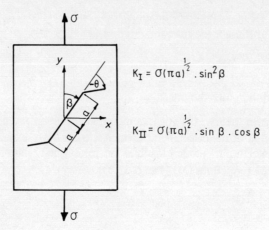

Fig. 5.2 *Angled crack in a plate under tension.*

$$K_I = \sigma(\pi a)^{\frac{1}{2}} \cdot \sin^2 \beta$$

$$K_{II} = \sigma(\pi a)^{\frac{1}{2}} \cdot \sin \beta \cdot \cos \beta$$

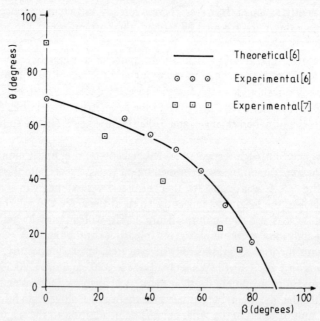

Fig. 5.3 *Fracture angle, θ, versus crack angle, β, in a cracked plate in tension.*

5.4 The strain energy density criterion (SEDC)

The strain energy density criterion attempts to account for both the onset of crack extension and the direction of crack extension under mixed-mode conditions [14–16]. It represents the first unified theory of fracture mechanics, and its success will depend upon experimental evidence.

The SED approach is based on the local value of the strain energy density, which is direction sensitive. Crack extension is postulated to occur in the direction of minimum strain energy density, when the strain energy density factor, S, attains a critical value S_c.

5.4.1 The strain energy density field for plane problems

The crack tip coordinate system is shown in Fig. 5.4.

The strain energy density in the vicinity of the crack tip is given, by [14]:

$$\frac{dW}{dV} = \frac{1}{\pi r}(a_{11}K_I^2 + 2a_{12}K_I K_{II} + a_{22}K_{II}^2 + a_{33}K_{III}^2) + \text{non-singular terms.} \quad (5.8)$$

Thus the strain energy density function possesses a $(1/r)$ singularity at the crack tip. Hence the expression

$$S = (a_{11}K_I^2 + 2a_{12}K_1 K_2 + a_{22}K_{II}^2 + a_{33}K_{III}^2)/\pi \quad (5.9)$$

represents the intensity of the strain energy density field in the vicinity of the crack tip. S is a function of the polar angle θ (Fig. 5.4) since, for plane strain:

$$a_{11} = \frac{1}{16G}[(3 - 4v - \cos\theta)(1 + \cos\theta)]$$

$$a_{12} = \frac{1}{16G} \cdot 2\sin\theta[\cos\theta - (1 - 2v)] \quad (5.10)$$

$$a_{22} = \frac{1}{16G}[4(1 - v)(1 - \cos\theta) + (1 + \cos\theta)(3\cos\theta - 1)]$$

$$a_{33} = 1/4G$$

where G is the shear modulus of elasticity and v is Poisson's ratio.

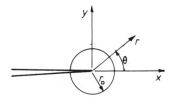

Fig. 5.4 *Crack tip coordinate system for strain energy density criterion.*

Thus, S describes the variation of the local energy density around the region where fracture will initiate, and is designated the strain energy density factor. It can be shown that S is invariant with reference to the K-coordinate system [14].

Attention is now confined to elements just outside the core region, radius r_0, since material properties in the immediate vicinity of the crack tip are unknown. Sih [15, 16] hypothesizes:

(a) Crack initiation is in the radial direction of minimum strain energy density, denoted θ_0, i.e.

$$\frac{dS}{d\theta} = 0, \quad \frac{d^2S}{d\theta^2} > 0 \quad \text{at } \theta = \theta_0 \tag{5.11}$$

(b) Crack extension occurs when the strain energy density factor S reaches a critical value S_c, i.e.

$$S_c = S(K_I, K_{II}, K_{III}) \quad \text{at } \theta = \theta_0. \tag{5.12}$$

The strain energy density factor S can be written as the sum of two components, one due to a change in volume (S_v), the other due to a change in shape (S_d), i.e.

$$S = S_v + S_d. \tag{5.13}$$

The coefficients a_{ij} in equation (5.9) may be regarded as the linear sum of two sets of components b_{ij} and $c_{ij}(i, j = 1, 2, 3)$ such that, Sih [17]:

$$S_v = (b_{11}K_I^2 + 2b_{12}K_IK_{II} + b_{22}K_{II}^2 + b_{33}K_{III}^2)/\pi \tag{5.14}$$

$$S_d = (c_{11}K_I^2 + 2c_{12}K_IK_{II} + c_{22}K_{II}^2 + c_{33}K_{III}^2)/\pi. \tag{5.15}$$

For plane strain the coefficients b_{ij}, c_{ij} may be written:

$$\begin{aligned}
b_{11} &= \frac{(1-2v)(1+v)}{12\mu}(1+\cos\theta) \\
b_{12} &= \frac{(1-2v)(1+v)}{12\mu}(-\sin\theta) \\
b_{22} &= \frac{(1-2v)(1+v)}{12\mu}(1-\cos\theta) \\
b_{33} &= 0
\end{aligned} \tag{5.16}$$

$$\begin{aligned}
c_{11} &= \frac{1}{16\mu}(1+\cos\theta)\left(\frac{2(1-2v)^2}{3} + 1 - \cos\theta\right) \\
c_{12} &= \frac{1}{16\mu}\sin\theta\left(2\cos\theta - \frac{2(1-2v)^2}{3}\right) \\
c_{22} &= \frac{1}{16\mu}\left(\frac{2(1-2v)^2}{3}(1-\cos\theta) - 4 - 3\sin^2\theta\right) \\
c_{33} &= 1/4\mu.
\end{aligned} \tag{5.17}$$

5.4.2 Physical significance of the strain energy density concept

Case 1: Mode I Loading

Consider an infinite sheet under uniaxial loading as shown in Fig. 5.5. By symmetry, $K_{II} = K_{III} = 0$. Thus from equations (5.9) and (5.10)

$$S = \frac{a_{11}}{\pi}K_I^2 = \frac{\sigma^2 a}{16G}[(3 - 4v - \cos\theta)(1 + \cos\theta)]. \tag{5.18}$$

There are two possible values for θ corresponding to $dS/d\theta = 0$ (designated θ_0).

Fig. 5.5 *Remote tension, mode I loading.*

(i) $\theta_0 = 0$ corresponds to S being a minimum, which occurs along $\theta = 0$, i.e.

$$S_{min} = \frac{(1 - 2v)}{4\mu}\sigma^2 a. \tag{5.19}$$

On separating S_{min} into $S_v | S_{min}$ and $S_d | S_{min}$ using equations (5.13), (5.14) and (5.15):

$$S_v | S_{min} = \frac{(1 + v)(1 - 2v)}{6\mu}\sigma^2 a \tag{5.20}$$

$$S_d | S_{min} = \frac{(1 - 2v)^2}{12\mu}\sigma^2 a \tag{5.21}$$

Thus, from (5.20) and (5.21)

$$S_v | S_{min} > S_d | S_{min}$$

which conforms with the concept that fracture occurs along a plane where $S_v > S_d$.

(ii) $\cos\theta_0 = 1 - 2v$ corresponds to S being a maximum, i.e.

$$S_{max} = \frac{(1 - v)^2}{4\mu}\sigma^2 a \tag{5.22}$$

Case 2: Combined mode I and mode II loading

Consider the angled crack in an infinite sheet under uniaxial tension or compression. Stress intensity factors for these configurations are given in Fig. 5.2. When these are combined with equations (5.10) and inserted into (5.9) the resulting expression for S may be differentiated with respect to θ to obtain stationary values of θ (designated θ_0), related to crack inclination β, by:

$$2(1 - 2v) \sin (\theta_0 - 2\beta) - \{2 \sin [2(\theta_0 - \beta)]\} - \sin^2 \theta_0 = 0. \quad (5.23)$$

Fracture angles predicted from equation (5.23) for the cases of uniaxial tension and compression are shown in Fig. 5.6.

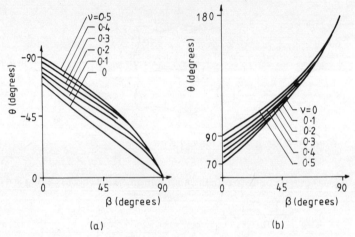

Fig. 5.6 *Crack angle, β, versus fracture angle, θ, strain energy density criterion prediction, plane strain (after [14]): (a) tension; (b) compression.*

Fast fracture prediction depends on a knowledge of S_c (equation (5.12)). For mode I loading it has been shown that, in the direction of fracture

$$S = \frac{K_I^2}{4G\pi} (1 - 2v) \quad (5.24)$$

thus, when $K_I \to K_{Ic}$

$$S_c = \frac{K_{Ic}^2}{4G\pi} (1 - 2v). \quad (5.25)$$

Therefore, S_c may be obtained from simple mode I tests of the standard type. With a knowledge of S_c it is possible to use the predicted fracture angles from equation (5.23) substituted in (5.9) to plot the mixed-mode fracture envelope (Fig. 5.7).

MIXED-MODE FRACTURE MECHANICS

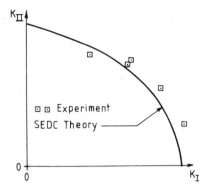

Fig. 5.7 *Mixed-mode fracture K_I versus K_{II}, strain energy density criterion prediction, plane strain (after [14]).*

5.5 Crack path stability

The standard technique for testing the stability of crack paths is to allow a slight kink or curve in the crack, and examine its progress thereafter. Cotterell and Rice [18] have produced a straightforward expression for crack tip stress intensity in this case by selecting an (x, y) coordinate system oriented parallel and perpendicular to the crack line at the crack tip (Fig. 5.8).

The stress intensities, to first order, are given by the expression:

$$\begin{Bmatrix} K_I \\ K_{II} \end{Bmatrix} = \left(\frac{1}{\pi a}\right)^{1/2} \int_0^{2a} \begin{Bmatrix} T_y(r) \\ T_x(r) \end{Bmatrix} \left(\frac{2a - r}{r}\right)^{1/2} dr$$

$$+ \left(\frac{a}{\pi}\right)^{1/2} \int_0^{2a} \begin{Bmatrix} T_x(r) \\ T_y(r) \end{Bmatrix} \left(\frac{\eta(2a)}{(2a)^2} - \frac{\eta(r)}{r^2}\right) \left(\frac{r}{2a - r}\right)^{1/2} dr \quad (5.26)$$

where T_x and T_y represent distributed tractions in x and y directions respectively, and $\eta(r)$ measures the deviation of the crack from the selected x-axis.

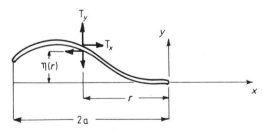

Fig. 5.8 *Coordinate system for kinked crack (after [18]).*

The first term is the result for a straight crack of length $2a$, and may be compared with equation (4.6), recognizing that $T_y \equiv \sigma_y$ and $T_x \equiv \tau_{xy}$, and shifting axes. The second term contains the effect of non-straightness of the crack.

By employing this solution, together with the assumption of a very small mode II component arising from loading device imperfections, the symmetric configuration illustrated in Fig. 5.9 was examined. The stress field ahead of the crack is also shown. $O(x^{1/2})$ indicates terms of order $x^{1/2}$ or greater. Note the

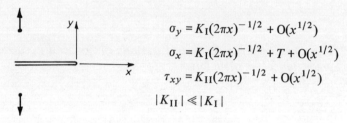

$$\sigma_y = K_\mathrm{I}(2\pi x)^{-1/2} + O(x^{1/2})$$
$$\sigma_x = K_\mathrm{I}(2\pi x)^{-1/2} + T + O(x^{1/2})$$
$$\tau_{xy} = K_\mathrm{II}(2\pi x)^{-1/2} + O(x^{1/2})$$
$$|K_\mathrm{II}| \ll |K_\mathrm{I}|$$

Fig. 5.9 *Symmetric loading configuration (after [18]).*

contribution to σ_x from T, a uniform tension acting parallel to the crack. Cotterell and Rice show that the stability of the crack path is determined entirely by the sign of T. When $T > 0$ the crack veers away from the crack plane (unstable), and when $T < 0$ the crack gradually returns to the initial crack line (stable) (see Fig. 5.10).

The above criterion appears to agree with experimental evidence. For instance, Radon, Lever and Culver [19] tested centrally cracked PMMA sheets under biaxial loading (Fig. 5.11). For this case $T = (\lambda - 1)\sigma$, and for all tests with

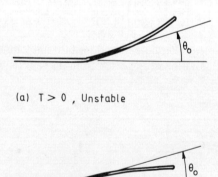

Fig. 5.10 *Kinked crack stability.*

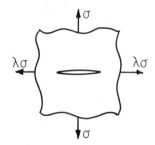

Fig. 5.11 *Biaxial loading configuration.*

$R < 1$ i.e. $T < 0$, the crack path was stable, whereas with $R > 1$, i.e. $T > 0$, the crack shows increasing deviation from the initial crack plane, the extent of this deviation increasing as λ increases over unity.

References

1. Johnson, R. E. (1973), *Apollo Experience Report – The Problem of Stress Corrosion Cracking*, NASA TN D-7111.
2. Pook, L. P. (1971), 'The effect of crack angle on fracture toughness', *Engng Frac. Mech.*, **3**, 205–18.
3. Beachem, C. D. (1968), 'Microscopic fracture processes' in *Fracture – An Advanced Treatise*, Vol. I, Ed. H. Liebowitz, Academic Press, New York.
4. Zackay, V. F., Gerberich, W. W. and Parker, E. R. (1968), 'Structural modes of fracture' in *Fracture – An Advanced Treatise*, Vol. I, Ed. H. Liebowitz, Academic Press, New York.
5. Broek, D. (1974), *Elementary Engineering Fracture Mechanics*, Noordhoff, Leiden.
6. Erdogan, F. and Sih, G. C. (1963), 'On crack extension in plates under plane loading and transverse shear', *Trans. ASME, J. Bas. Engng*, **85**, 519–27.
7. Wilson, W. K. (1969), *On Combined Mode Fracture Mechanics*, Westinghouse Research Labs, Report 69–1E7–F MECH–R1.
8. Tanaka, K. (1974), 'Fatigue crack propagation from a crack inclined to the cyclic tensile axis', *Engng Frac. Mech.*, **6**, 493–507.
9. Wu, E. M. (1967), 'Application of fracture mechanics to anisotropic plates', *Trans ASME, J. Appl. Mech.*, **34**, 967–74.
10. Hoskin, B. C., Graff, D. G. and Foden, P. J. (1965), *Fracture of Tension Panels with Oblique Cracks*, Aer. Res. Labs, Melbourne, Report No. SM 305.
11. Irwin, G. R. (1957), 'Analysis of stresses and strains near the end of a crack traversing a plate', *J. Appl. Mech.*, **24**, 361–4.
12. Williams, J. G. and Ewing, P. D. (1972), 'Fracture under complex stress – the angled crack problem', *Int. J. Frac. Mech.*, **8**, No. 4, 441–5.

13. Cotterell, B. (1966), 'The influence of the stress distribution at the tip of a crack', *Int. J. Frac. Mech.*, **2**, No. 3, 526–33.
14. Sih, G. C. (1973), 'Methods of analysis and solutions of crack problems', *Mechanics of Fracture*, Vol. I, Ed. G. C. Sih, Noordhoff, Leiden.
15. Sih, G. C. (1973), *Handbook of Stress Intensity Factors for Researchers and Engineers*, Institute of Fracture and Solid Mechanics, Lehigh University, Bethlehem, Pennsylvania.
16. Sih, G. C. (1973), 'Some basic problems in fracture mechanics and new concepts', *Engng Frac. Mech.*, **5**, 365–77.
17. Sih, G. C. (1974), 'Strain energy density factor applied to mixed mode crack problems', *Int. J. Frac.*, **10**, No. 3, 305–21.
18. Rice, J. R. (1979), 'The mechanics of quasi-static crack growth', *Proc. 8th US Natl Congr. Appl. Mech., June 1978*, Ed. R. E. Kelly, Western Periodicals Co., Hollywood, California.
19. Radon, J. C., Lever, P. S. and Culver, L. E. (1977), 'Fracture toughness of PMMA under biaxial stress' in *Fracture 1977 Proc. 4th Int Conf. on Fracture*, Ed. D. M. R. Taplin, Vol. 3, pp. 1113–18, University of Waterloo Press.

6 Crack Tip Plasticity and Associated Effects

6.1 Introduction

Thus far we have implicitly assumed that the material behaves in a purely elastic manner. In practice most materials deform plastically once some critical combination of stresses is achieved. This may require modification of some of the linear elastic fracture mechanics concepts. The simplest yield criterion assumes that a material will fail plastically when the maximum principal stress reaches the uniaxial yield stress of the material (this criterion is in fact not fully acceptable on physical grounds, nevertheless it is used for some simple models).

The two most common, physically acceptable yield criteria, are due to Tresca and Von Mises [1, 2]. The Tresca criterion predicts that yielding will occur when the maximum value of shear stress reaches a critical value. When written in terms of the principal stresses $\sigma_1, \sigma_2, \sigma_3$ ($\sigma_1 > \sigma_2 > \sigma_3$) and the uniaxial yield stress, Y, Tresca's criterion predicts yielding when:

$$|\sigma_1 - \sigma_3| = Y. \tag{6.1}$$

Von Mises' criterion requires the shear strain energy per unit volume to reach a critical value. This criterion, expressed in terms of principal stresses and uniaxial yield stress predicts failure when:

$$(\sigma_1 - \sigma_2)^2 + (\sigma_2 - \sigma_3)^2 + (\sigma_3 - \sigma_1)^2 = 2Y^2. \tag{6.2}$$

The two criteria are in agreement for certain stress combinations, but differ for others. Von Mises' generally produces better agreement with experimental observations, whilst the Tresca criterion is frequently easier to use in practice.

6.2 Irwin's plastic zone model

Consider the variation of σ_y along $y = 0$ for a remotely loaded cracked sheet in a state of plane stress (Fig. 6.1(a)). Assuming the simplest yield criterion, σ_y will exceed the material's yield stress, Y, at some distance from the crack tip, r^*. If we take r^* as the first estimate of the extent of the plastic zone, it is clear that the force produced by the stress shown shaded in Fig. 6.1(a), acting over length r^* will produce further yielding. In fact the whole stress curve must be shifted as shown in Fig. 6.1(b), so that equilibrium is maintained, resulting in a plastic

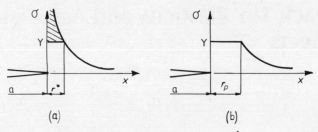

Fig. 6.1 (a) *Elastic stress distribution and first estimate of plastic zone size;* (b) *second estimate of plastic zone size.*

zone of length r_p. Recalling the expression for stresses in the near-tip region (Section 3.3), we may state the equilibrium condition as:

$$\int_0^{r^*} \frac{K}{(2\pi r)^{1/2}} \, dr = r_p Y. \tag{6.3}$$

Thus:

$$\left(\frac{2}{\pi}\right)^{1/2} K r^{*1/2} = r_p Y. \tag{6.4}$$

At $r = r^*$, $\sigma_y = K/(2\pi r^*)^{1/2} = Y$, substituting into (6.4) gives the revised estimate of plastic zone size, r_p, as:

$$r_p = \frac{1}{\pi}\left(\frac{K}{Y}\right)^2. \tag{6.5}$$

If at this stage we visualize the plastic zone as being circular, it will have radius of $r_p/2$, designated λ, thus:

$$\lambda = \frac{1}{2\pi}\left(\frac{K}{Y}\right)^2. \tag{6.6}$$

Irwin [3] proposed a plastic zone correction, based on a 'notional' crack which extends to the centre of the plastic zone. Thus the crack behaves as if it were of length $(a + \lambda)$ (Fig. 6.2).

In order to produce a consistent result, we must also associate a new stress intensity factor, K^*, with this notional crack, such that:

$$K^* = Q\sigma[\pi(a + \lambda)]^{1/2}. \tag{6.7}$$

We have derived an expression for crack-opening, v, in Section 3.5. For the case of plane stress this becomes:

$$v = \frac{2\sigma}{E}(a^2 - x^2)^{1/2}. \tag{6.8}$$

CRACK TIP PLASTICITY AND ASSOCIATED EFFECTS 103

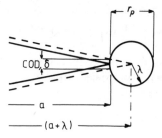

Fig. 6.2 *Irwin's plastic zone correction.*

In the absence of plasticity, $v = 0$ at $x = a$, the crack tip. However, considering the notional crack of length $(a + \lambda)$

$$v = \frac{2\sigma}{E}[(a+\lambda)^2 - x^2]^{1/2} \tag{6.9}$$

and at $x = a$, the crack-opening (termed the crack tip opening displacement (COD), designated δ) is given by $2v$ (see Fig. 6.2). Neglecting small terms:

$$\delta = \frac{4\sigma}{E}(2a\lambda)^{1/2}. \tag{6.10}$$

Substituting from (6.6) we find:

$$\delta = \frac{4}{\pi}\frac{K^2}{EY}. \tag{6.11}$$

6.3 Dugdale's plastic zone model

An alternative approach to that of Irwin was proposed by Dugdale [4], and in equivalent form by Barenblatt [5]. In this case the crack is assumed to extend right through the plastic zone (Fig. 6.3), thus the crack length is increased from $2a$ to $2(a + \rho)$, where ρ is the extent of the plastic zone. Furthermore, that

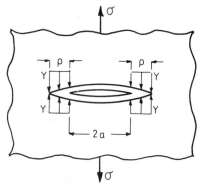

Fig. 6.3 *Dugdale crack and plastic zones.*

proportion of the extended crack situated in the plastic zone experiences a constant, negative pressure of the same magnitude as the yield stress.

Dugdale postulated that the plastic zone dimension, ρ, is fixed by the requirement that the stress singularity should disappear, thus the superposition of stress intensity due to remotely applied stress, K_σ, and that due to plastic closure, K_Y, should be zero, giving:

$$K_\sigma + K_Y = 0. \tag{6.12}$$

Now we know that:

$$K_\sigma = \sigma[\pi(a+\rho)]^{1/2} \tag{6.13}$$

and we can obtain K_Y from the weight function via equation (4.13), in this case $p^*(x)$ consists of a negative pressure of magnitude Y extending from a to $(a+\rho)$, and the integral becomes:

$$K_Y = -2\int_a^{a+\rho} Ym(x,(a+\rho))\,\mathrm{d}(a+\rho) \tag{6.14}$$

where the minus sign indicates that the loading tends to produce crack closure. This may be solved in a straightforward manner, giving:

$$K_Y = -2Y\left(\frac{a+\rho}{\pi}\right)^{1/2}\arccos\left(\frac{a}{a+\rho}\right). \tag{6.15}$$

Substituting (6.13) and (6.15) into (6.12) we obtain:

$$\frac{a}{a+\rho} = \cos\left(\frac{\pi\sigma}{2Y}\right). \tag{6.16}$$

At low remote stress levels, $\sigma \ll Y$, the series expansion of the right-hand side of the above equation gives:

$$\rho = \frac{\pi a}{8}\left(\frac{\sigma}{Y}\right)^2 = \frac{\pi}{8}\left(\frac{K}{Y}\right)^2. \tag{6.17}$$

This may be compared with Irwin's expression (equation (6.5)), and general agreement is noted. When σ is no longer small compared with Y, it is necessary to employ equation (6.16) to determine ρ.

It is also possible to employ the weight function in the determination of crack shape, along the lines of Example 4.1. We shall derive separately the crack shape appropriate to the remote and plastic zone loads, and calculate the final crack shape by superposition. Recalling the relationship between crack shape and weight function (equation (4.27)), for a crack of half-length $(a+\rho)$, under plane stress:

$$\frac{\partial v}{\partial a} = \frac{2K}{E} m(x,(a+\rho)) \tag{6.18}$$

CRACK TIP PLASTICITY AND ASSOCIATED EFFECTS 105

where the weight function $m(x, (a + \rho))$ is:

$$m(x, (a + \rho)) = \left(\frac{a+\rho}{\pi}\right)^{1/2} [(a+\rho)^2 - x^2]^{-1/2}. \tag{6.19}$$

Remote loading
In this case $K_\sigma = \sigma[\pi(a + \rho)]^{1/2}$, and the weight function is given as (6.19). Manipulation as in Example 4.1 yields the crack shape:

$$v_\sigma = \frac{2\sigma}{E}[(a+\rho)^2 - x^2]^{1/2}. \tag{6.20}$$

Re-arranging equation (6.16) we find:

$$\sigma = \frac{2Y}{\pi}\arccos\left(\frac{a}{a+\rho}\right) \tag{6.21}$$

which may be substituted into (6.20) to obtain:

$$v_\sigma = \frac{4Y}{\pi E}\left[\arccos\left(\frac{a}{a+\rho}\right)\right][(a+\rho)^2 - x^2]^{1/2}. \tag{6.22}$$

Plastic zone loading
In this case K_Y is given in equation (6.15), and the weight function by equation (6.19). Substitution into (6.18) gives:

$$\frac{\partial v_Y}{\partial(a+\rho)} = -\frac{4Y}{\pi E}\arccos\left(\frac{a}{a+\rho}\right)\frac{a+\rho}{[(a+\rho)^2 - x^2]^{1/2}} \tag{6.23}$$

Integrating by parts we obtain:

$$v_Y = -\frac{4Y}{\pi E}\left[\arccos\left(\frac{a}{a+\rho}\right)[(a+\rho)^2 - x^2]^{1/2}\right.$$
$$\left. - \int \frac{[(a+\rho)^2 - x^2]^{1/2}}{[(a+\rho)^2 - a^2]^{1/2}}\left(\frac{a}{a+\rho}\right)d(a+\rho)\right]. \tag{6.24}$$

At this stage we can superimpose the two displacement expressions by adding (6.22) and (6.24). With the equal and opposite terms cancelling, this produces the total displacement, v:

$$v = v_\sigma + v_Y = \frac{4Y}{\pi E}\int\frac{[(a+\rho)^2 - x^2]^{1/2}}{[(a+\rho)^2 - a^2]^{1/2}}\left(\frac{a}{a+\rho}\right)d(a+\rho). \tag{6.25}$$

This integration is straightforward, if somewhat tedious, by using the substitution:

$$t^2 = \frac{(a+\rho)^2 - x^2}{(a+\rho)^2 - a^2}. \tag{6.26}$$

We obtain:

$$v = \frac{4Y}{\pi E}\left[\frac{a}{2}\ln\left(\frac{t+1}{t-1}\right) - \frac{x}{2}\ln\left(\frac{t+(x/a)}{t-(x/a)}\right)\right], \qquad (6.27)$$

the constant of integration being zero as a result of the requirement $v = 0$ at $x = (a + \rho)$. The above result is equivalent to that of [6], derived via Westergaard stress functions.

If we wish to know the COD, δ, it is necessary to evaluate $2v$ as $x \to a$. Using L'Hôpital's rule for the limit:

$$\delta = 2 \lim_{x \to a} v = \frac{8}{\pi}\frac{Y}{E} a \ln\left(\frac{a+\rho}{a}\right). \qquad (6.28)$$

6.4 Plastic zone shapes

The previous models were based on an oversimplified yield criterion and plastic zone shape. In order to apply a proper yield criterion we must rewrite the crack tip stress field equations (3.37) in terms of principal stresses (see Section 1.22) giving:

$$\sigma_1 = \frac{K}{(2\pi r)^{1/2}}\cos\left(\frac{\theta}{2}\right)\left[1 + \sin\left(\frac{\theta}{2}\right)\right]$$

$$\sigma_2 = \frac{K}{(2\pi r)^{1/2}}\cos\left(\frac{\theta}{2}\right)\left[1 - \sin\left(\frac{\theta}{2}\right)\right] \qquad (6.29)$$

$$\sigma_3 = \begin{cases} v(\sigma_1 + \sigma_2) = 2v\dfrac{K}{(2\pi r)^{1/2}}\cos\left(\dfrac{\theta}{2}\right), & \text{plane strain} \\ 0, & \text{plane stress.} \end{cases}$$

By substituting the above relationship into Von Mises' criterion (6.2) we obtain an expression for the plastic zone boundary as a function of θ, $r_p(\theta)$. This produces, for the case of plane strain:

$$\frac{K^2}{2\pi r_p}\left[\tfrac{3}{2}\sin^2\theta + (1-2v)^2(1+\cos\theta)\right] = 2Y^2 \qquad (6.30)$$

whilst for plane stress:

$$\frac{K^2}{2\pi r_p}(1 + \tfrac{3}{2}\sin^2\theta + \cos\theta) = 2Y^2. \qquad (6.31)$$

The above equations may be re-arranged to obtain:

$$r_p(\theta) = \begin{cases} \dfrac{K^2}{4\pi Y^2}\left[\tfrac{3}{2}\sin^2\theta + (1-2v)^2(1+\cos\theta)\right], & \text{plane strain} \quad (6.32) \\ \dfrac{K^2}{4\pi Y^2}\left[1 + \tfrac{3}{2}\sin^2\theta + \cos\theta\right], & \text{plane stress.} \quad (6.33) \end{cases}$$

CRACK TIP PLASTICITY AND ASSOCIATED EFFECTS

The form of the plastic zone boundaries predicted from (6.32) and (6.33) is shown in Fig. 6.4. Notice that the plane strain plastic zone is smaller and, for certain values of θ, very much smaller, than the equivalent plane stress zone.

Derivation of plastic zone dimensions via the Tresca criterion (6.1) is straightforward. The reader may wish to derive the relationships:

$$r_p = \frac{K^2}{2\pi Y^2}\left\{\cos\left(\frac{\theta}{2}\right)\left[1 + \sin\left(\frac{\theta}{2}\right)\right]\right\}^2 \qquad \text{Plane stress} \quad (6.34)$$

or

$$\left. \begin{array}{l} r_p = \dfrac{K^2}{2\pi Y^2}\cos^2\left(\dfrac{\theta}{2}\right)\left[1 - 2\nu + \sin\left(\dfrac{\theta}{2}\right)\right]^2 \\[1em] r_p = \dfrac{K^2}{2\pi Y^2}\cos^2\left(\dfrac{\theta}{2}\right) \end{array} \right\} \quad \text{plane strain} \quad (6.35)$$

whichever is the greater.

The above results are also illustrated in Fig. 6.4. The dimensions of the Tresca criterion zone is invariably greater than, or equal to that of the Von Mises' zone. By substituting $\theta = 0$ into equations (6.33) or (6.35) we obtain a value of r_p which is half of that given in equation (6.5). However, the latter equation was corrected to account for the redistribution of stress outside the assumed elastic/plastic boundary. The analytical and experimental details of the full plastic zone correction are somewhat complicated, and indeed still under investigation. We simply note that a recent theoretical study due to Shih [7], which includes the effects of strain hardening, appears to be confirmed by experimental work due to Hahn and Rosenfield [8].

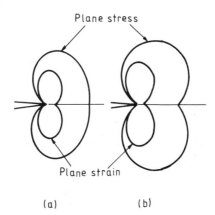

Fig. 6.4 *Plastic zone shapes:* (a) *Von Mises' criterion;* (b) *Tresca's criterion (after [14]).*

6.5 Types of failures

Metals and other materials can break into separate pieces in several ways. The conventional tensile test produces a variety of failure modes, depending upon material, environment and loading conditions. Soft metals tend to neck and thin to a fine point before separation takes place, rather like the result of slowly extending a piece of plasticine (Fig. 6.5(a)).

At the other extreme are very brittle materials such as glass and hardened steel, with very little deformation before failure (Fig. 6.5(b)), which produce a macroscopically flat fracture surface. However, by far the most common types of failure are those which exhibit some necking or deformation, finally breaking to produce a characteristic 'cup and cone' fracture surface (Fig. 6.5(c)).

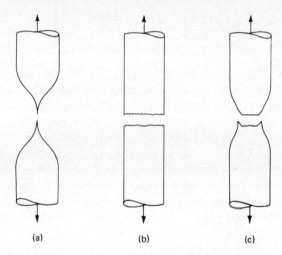

Fig. 6.5 *Types of fracture:* (a) *point;* (b) *flat;* (c) *cup and cone.*

A similar behaviour is observed in the fracture of moderately thick plate with through-the-thickness cracks. Fig. 6.6 illustrates schematically the types of fracture surface, and the variation of experimentally measured fracture toughness, K_c (the critical stress intensity factor at failure), with plate thickness. Clearly, the critical stress intensity factor attains a minimum value and becomes geometry independent only when the plate is sufficiently thick. This minimum value of K_c is designated K_{Ic}, the plane strain fracture toughness. The fracture surface when the K_{Ic} value is reached is almost totally macroscopically flat. The dotted curve in Fig. 6.6 indicates the proportion of flat fracture surface at different plate thicknesses.

In order to understand the reasons for this apparently complex failure mechanism, consider two small rectangular elements of material, one in the centre

CRACK TIP PLASTICITY AND ASSOCIATED EFFECTS 109

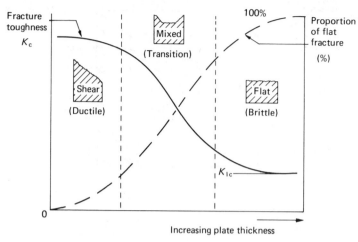

Fig. 6.6 *Variation in K_c, amount of flat fracture and failure mode with plate thickness.*

of a moderately thick plate, close to the crack tip, the other in a similar position relative to the crack, but at the free surface of the plate normal to the crack front (Fig. 6.7). As the remote loading is increased, each of these elements will fail at some particular load level, either by shear (sliding of one plane of atoms over another, governed by Tresca's or Von Mises' criteria which were mentioned in Section 6.1), or by cleavage (direct separation of one plane of atoms from another caused by loading normal to the eventual fracture surface). Inspection of the yield criteria indicates that a hydrostatic stress state ($\sigma_1 = \sigma_2 = \sigma_3$) cannot produce a ductile failure. Thus, when a load level is reached which will cause ductile failure of the plane stress element (A) on planes at ±45 degrees to the crack plane as illustrated in Fig. 6.7, the near-hydrostatic system experienced by the plane strain element (B) will not be capable of producing ductile failure until a much greater load is applied. In this case element B may fail by cleavage in the plane of the crack before it is able to achieve a critical shear stress level. Thus we associate the slant (shear lip) failure with a ductile failure on inclined planes, and the flat (crack plane) failure with cleavage separation.

As the crack grows, the plastic zone ahead of the crack tip also becomes larger, allowing less through-the-thickness restraint on internal elements, and conditions approach plane stress right through the thickness. Under these conditions the proportion of flat fracture surface reduces as the crack extends, and the proportion of slant (shear lip) increases. The resulting failure is illustrated in Fig. 6.8.

In a real, cracked specimen which is thick enough to produce sufficient constraint, and hence plane strain conditions in the interior, conditions of plane stress will still prevail at the surface, since it is free of stresses. As a result the

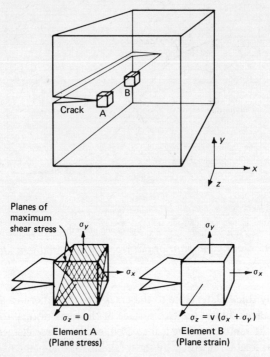

Fig. 6.7 *Plane stress and plane strain elements near the crack tip.*

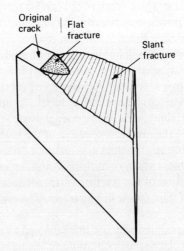

Fig. 6.8 *Final fracture surface of moderately thick plate.*

CRACK TIP PLASTICITY AND ASSOCIATED EFFECTS

shape and size of the plastic zone changes from that appropriate to plane strain in the interior, to that of plane stress at the surface. This effect is shown in Fig. 6.9.

Furthermore, if the loading is increased, so is the size of the plastic zone, which tends to reduce the through-the-thickness constraint, allowing more of the plastic zone to deviate from the plane strain condition. Thus large plastic zones tend to allow plane stress conditions to develop.

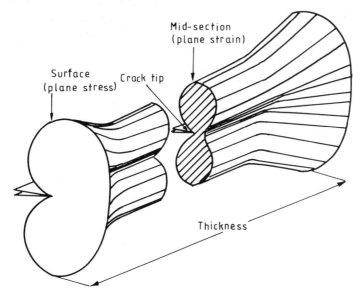

Fig. 6.9 *Variation of crack tip plastic zone through the thickness of a plate.*

The ratio of plastic zone size to specimen thickness must be less than unity for plane strain to predominate. Since it is the plane strain condition which produces the lower limiting value of fracture toughness, very thick plates may be required for plane strain fracture toughness testing, particularly if the material has high K_{Ic} and low Y (see equation (6.5)).

6.6 R-curves

Recalling our brief look at energy concepts in Chapter 2, we concluded that failure occurs when the total energy of the system is increased as a result of an incremental crack extension, thus failure occurs when:

$$G = R$$

where G is the energy release rate (dU/da), and R is the crack resistance (dW/da), sometimes termed crack resistance force since it has the units of force.

In Chapter 2 we assumed that R was independent of crack length, since dW/da is shown constant in Fig. 2.3. This is approximately true for cracks under plane strain conditions, however in situations involving larger proportions of plane stress (slant) failure, R is no longer independent of crack length, and fracture toughness evaluations are frequently made using R-curve (or resistance curve) methods.

R-curves characterize fracture resistance during incremental, stable crack extension. They indicate the variation of toughness as the crack is driven stably by a gradually increasing load. Such fracture resistance will depend upon specimen thickness as well as test temperature and rate of application of load. Referring to Fig. 6.10, when a specimen is loaded to a stress σ_i the crack starts to propagate. However, this growth is stable, and fracture does not occur. Maintaining σ_i the crack propagates only a small distance and then stops. Further stress increments produce more stable crack growth, until at some critical stress, σ_c a critical crack size, a_c is achieved.

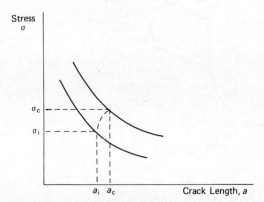

Fig. 6.10 *Stable crack growth (after [17]).*

Fig. 6.11 illustrates schematically a typical R-curve, showing fracture resistance R as a function of effective crack length. The R value is calculated by using the effective crack length, and is plotted against the actual crack extension. Stable crack propagation occurs under increasing load until the point of tangency, G_c between the R-curve and the crack driving force G-curve is reached. At this stage crack propagation becomes unstable. For an internal crack in an infinite sheet under remote tension the curve of G versus crack length would be a straight line through the origin. Whilst R-curves may be determined experimentally [9], [10] and analytically [11], the techniques and concepts are not yet fully developed.

CRACK TIP PLASTICITY AND ASSOCIATED EFFECTS

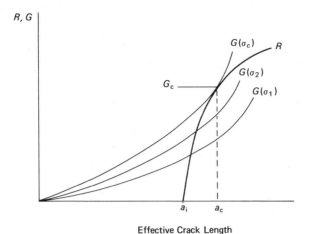

Fig. 6.11 *Typical R-curve.*

6.7 Plane strain fracture toughness testing

Let us confine our attention to the determination of the lower limiting value of fracture toughness, the plane strain fracture toughness, K_{Ic}. The requirements of the testing procedure are:

(a) The test must be a valid one.
(b) The specimens should be standardized, and relatively easy to produce.
(c) The specimens should be economical in use of material.
(d) The testing procedure should be straightforward, and the results reproducible.

The two most common specimen types currently employed in mode 1 testing both include single edge notches, which are extended by fatigue precracking prior to testing. The single edge-cracked bend specimen (Fig. 6.12(a)) is subjected to three-point bending, whilst the compact tension specimen (Fig. 6.12(b)) is pin-loaded in order to produce combined tension and bending.

Clearly, stress intensity factor, K, is required for each specimen type, to a high accuracy. A combination of numerical methods and experimental verification has confirmed the standard relationships given below [12]:

Bend specimen

$$K = \frac{3PL}{BW^{3/2}} \left[1.93 \left(\frac{a}{W}\right)^{1/2} - 3.07 \left(\frac{a}{W}\right)^{3/2} + 14.53 \left(\frac{a}{W}\right)^{5/2} - 25.11 \left(\frac{a}{W}\right)^{7/2} + 25.80 \left(\frac{a}{W}\right)^{9/2} \right] \quad (6.36)$$

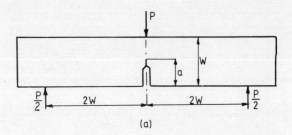

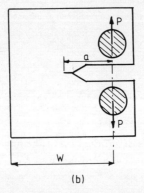

Fig. 6.12 *Standard fracture toughness testing specimens:* (a) *single edge-cracked bend specimen;* (b) *compact tension specimen (after [9]).*

where $L = 2W$, thus

$$K = \frac{P}{BW^{1/2}} Y_1 \left(\frac{a}{W}\right).$$

Alternatively, Y_1 may be derived directly from Table 6.1.

Table 6.1

$\frac{a}{W}$	Y_1	$\frac{a}{W}$	Y_1	$\frac{a}{W}$	Y_1	$\frac{a}{W}$	Y_1
0.450	9.1	0.455	9.23	0.510	10.96	0.515	11.14
0.460	9.37	0.465	9.52	0.520	11.33	0.525	11.52
0.470	9.66	0.475	9.81	0.530	11.71	0.535	11.91
0.480	9.96	0.485	10.12	0.540	12.12	0.545	12.33
0.490	10.28	0.495	10.44	0.550	12.55		—
0.500	10.61	0.505	10.78				

Compact tension specimen

$$K = \frac{P}{BW^{1/2}} \left[29.6 \left(\frac{a}{W}\right)^{1/2} - 185.5 \left(\frac{a}{W}\right)^{3/2} + 655.7 \left(\frac{a}{W}\right)^{5/2} - 1017 \left(\frac{a}{W}\right)^{7/2} + 638.9 \left(\frac{a}{W}\right)^{9/2} \right]. \quad (6.37)$$

In this case

$$K = \frac{P}{BW^{1/2}} Y_2 \left(\frac{a}{W}\right).$$

Y_2 is given in Table 6.2.

Table 6.2

$\frac{a}{W}$	Y_2	$\frac{a}{W}$	Y_2	$\frac{a}{W}$	Y_2	$\frac{a}{W}$	Y_2
0.450	8.34	0.455	8.45	0.510	9.90	0.515	10.05
0.460	8.57	0.465	8.69	0.520	10.21	0.525	10.37
0.470	8.81	0.475	8.93	0.530	10.54	0.535	10.71
0.480	9.06	0.485	9.19	0.540	10.89	0.545	11.07
0.490	9.32	0.495	9.46	0.550	11.26		
0.500	9.60	0.505	9.75				

The specimen dimensions must be sufficiently large in comparison with the plastic zone dimensions. The three relevant dimensions are crack length, a, specimen thickness, B, and uncracked ligament length, $(W - a)$, the following size requirements are normally specified [12, 13]:

$$a, B, W/2 > 2.5 \left(\frac{K_{Ic}}{Y}\right)^2. \quad (6.38)$$

Here is an interesting paradox, in order to determine K_{Ic} it is necessary to specify the specimen dimensions on the basis of a known K_{Ic}! However, by making a suitable overestimate of K_{Ic} on the basis of known values for similar materials, and checking the validity after the test, valid results will normally be obtained. Subsequent tests may then make use of more economically dimensioned specimens.

Next, drawings are produced, and the test specimens machined in accordance with the relevant standard. Remember that the crack length, a, includes both the notch and the precracking. The latter is produced by cyclically loading and unloading the specimen, the maximum stress intensity level being significantly less than K_{Ic} in order to avoid the effects of residual stress fields along the crack

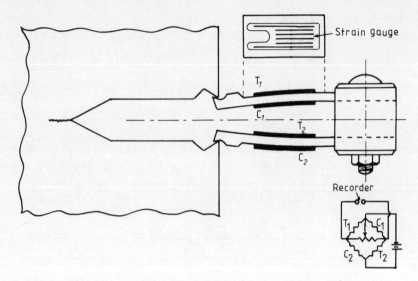

Fig. 6.13 *Clip gauge and electrical bridge arrangement (after [9]).*

line. This method of precracking produces a sufficiently sharp crack for valid testing.

For the test itself a clip gauge (Fig. 6.13) is mounted across the crack mouth, and a measure of crack opening is obtained via strain gauges fixed to the surfaces of the gauge arms. The load, P, on the specimen is increased until fast fracture occurs, as indicated by a gross non-linearity in the load–displacement record (Fig. 6.14). The procedure by which the exact load at fast fracture, P_Q, is determined is clearly explained in the standards [12, 13].

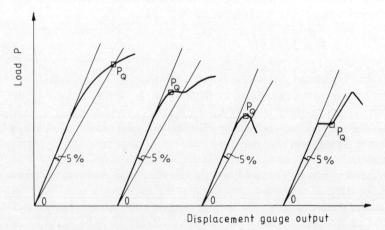

Fig. 6.14 *Typical load-displacement record for test specimens (after [9]).*

CRACK TIP PLASTICITY AND ASSOCIATED EFFECTS

Finally, the stress intensity factor at failure, K_Q, may be determined via equations (6.36) or (6.37), and provided the size criteria of equation (6.38) are satisfied, K_Q may be accepted as a valid plane strain fracture toughness.

6.8 Failure criteria in the presence of moderate plasticity

We may use standard test procedures to determine values of the plane strain fracture toughness, K_{Ic}. The tougher the material, the larger must be the specimen in order to meet the criteria laid down for validity of the test. These ensure that the plastic zone dimensions are small compared with other dimensions of the test piece.

Clearly it would be advantageous if one were able to conduct such tests using smaller specimens. Furthermore, we wish to be able to predict failure in practical situations involving significant amounts of plasticity. We now look at two proposed criteria which are intended for use in such situations. The first is based on the critical value of crack tip opening, the second on a critical value of strain energy release rate with non-linear elastic behaviour.

6.8.1 *Crack tip opening displacement (COD)*

Wells [14] proposed that the failure of cracked specimens in the presence of moderate plasticity could be characterized by the crack tip opening displacement (COD) which was defined in Section 6.2. Referring to the Dugdale model the value of COD, δ, may be written, after substituting from (6.21) into (6.28):

$$\delta = \frac{8Y}{\pi E} a \ln\left[\sec\left(\frac{\pi\sigma}{2Y}\right)\right]. \tag{6.39}$$

expanding this in series form:

$$\delta = \frac{8Y}{\pi E} a \left[\frac{1}{2}\left(\frac{\pi\sigma}{2Y}\right)^2 + \frac{1}{12}\left(\frac{\pi\sigma}{2Y}\right)^4 + \frac{1}{45}\left(\frac{\pi\sigma}{2Y}\right)^6 + \cdots\right] \tag{6.40}$$

For remote load $\sigma < 0.7Y$, a reasonable approximation to δ is:

$$\delta = \pi\sigma^2 a/EY. \tag{6.41}$$

But $K = \sigma(\pi a)^{1/2}$, hence:

$$\delta = K^2/YE. \tag{6.42}$$

In general, at crack instability the above expression becomes [15]:

$$\delta_c = \frac{K_{Ic}^2(1-\nu^2)}{\lambda EY} \tag{6.43}$$

where $(1-\nu^2)$ may be set to unity for plane stress conditions, and λ is a constant constraint factor. Since the right-hand side of equation (6.43) is a function of critical crack length for a particular configuration, we may assume that the same is true of δ_c, the critical value of COD. The advantage of the latter formulation is that COD values may be measured on relatively small specimens, with moderate plasticity. For further details of test method and results the reader is referred to [16], [17] or [18].

6.8.2 *The J integral*

In section 2.3.4 we considered the strain energy release rate for a small crack extension on the basis of assumed linear elastic behaviour. If the plastic zone dimensions are no longer negligibly small, the energy release rate, G, will not be valid.

By taking a line integral along a contour Γ surrounding the crack tip, starting on the lower crack surface, and moving anticlockwise to the upper surface (Fig. 6.15), we may define J, where:

$$J = \int_\Gamma W \, dy - T \cdot \left(\frac{\partial u}{\partial x}\right) ds, \qquad (6.44)$$

s is the arc length, T the outward traction vector on Γ, and u the displacement vector.

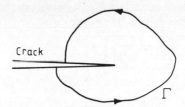

Fig. 6.15 *Unclosed contour around crack tip for J integral.*

Rice [19] has shown that in the case of a closed contour, $J = 0$. Thus, the J integral taken around the closed contour ABCDEF shown in Fig. 6.16 is zero. Since no contribution arises from CD and AF on the crack surfaces (i.e. $T = 0$ and $dy = 0$), the integral along ABC must be equal and opposite to that along DEF. Therefore, the J integral taken along an unclosed contour between unloaded crack surfaces is *path independent*.

Evaluation of (6.44) for the case of linear elasticity leads to [19]:

$$J = G \qquad (6.45)$$

whilst, in general

$$J = -\partial V/\partial a \qquad (6.46)$$

CRACK TIP PLASTICITY AND ASSOCIATED EFFECTS

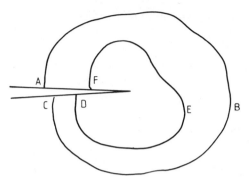

Fig. 6.16 *Closed contour at crack tip.*

where V is the potential energy per unit thickness. For the case of linear elasticity reference to Section 2.3.4 indicates that equations (6.45) and (6.46) are equivalent.

Thus J characterizes energy release rate during a small crack extension, which may also be valid in the presence of significant crack tip plasticity. It is reasonable to expect that there will be a critical value of J, termed J_{Ic}, at which crack extension occurs. Since this must also hold for the purely elastic case, it follows that:

$$J_{Ic} = G_{Ic} \tag{6.47}$$

implying that a failure criterion associated with significant plasticity can be used to determine G_{Ic} and vice versa. Since J is path independent we may choose the most convenient path, usually the specimen boundary. Equation (6.46) suggests a physical interpretation of J. By testing a specimen of the type illustrated in Fig. 6.17(a), at crack length a and $(a + \delta a)$, a load-displacement record

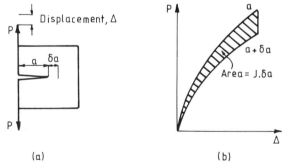

Fig. 6.17 *Physical interpretation of the J integral:* (a) *typical experimental arrangement;* (b) *load–displacement record at crack lengths a and* $(a + \delta a)$.

illustrated in Fig. 6.17(b) is obtained. The shaded region is

$$\partial U = J\,\delta a$$

the displacement Δ being measured at the loading points.

Thus J is seen to be the potential energy difference between two bodies having identical boundary tractions, and incrementally different crack lengths.

6.9 A note on general yield loads

Linear elastic fracture mechanics (LEFM) can be applied to problems in which the plastic zone is small compared to the crack length. Fracture is then characterized by the plane strain fracture toughness, K_{Ic}. In circumstances dominated by plane stress, the plastic zone is larger; nevertheless, if failure still occurs at net section stress levels which are much lower than the yield stress, problems may still be tractable by LEFM based methods.

However, when the plastic zone is large in comparison with the crack size, LEFM can no longer be used. In such cases the plastic zone spreads through the whole of the net section. This will occur in the case of low and medium toughness materials with very short cracks, and in very high toughness materials with moderate length cracks. In each case the explanation is simply that the crack tip stress intensity factor cannot reach a critical value before net section yielding occurs.

Whilst it is not the aim of this text to examine fully plastic failure, we simply note that in general the net section stress at plastic failure, σ_{net} is given by

$$\sigma_{net} = \gamma Y$$

where Y is the uniaxial yield stress for the material, and γ is some constraint factor dependent upon the geometry of the specimen, the determination of which will involve the use of slip line field theory.

The reader may also wish to note that the concepts of R curve, J integral and slip line fields are all incorporated in recent work on elastic–plastic crack growth [20].

6.10 The failure assessment diagram

Some recent work has led to the development of a so-called failure assessment diagram (FAD). The FAD presents a single curve which represents a 'safe envelope'. Provided a component is loaded so as to represent a point inside the envelope it is safe from failure, whether fully elastic, fully plastic or a combination of modes. The position of the point relative to the failure envelope is an indication of the factor of safety under the particular loading conditions.

It is claimed that the FAD not only reduces difficult concepts to an easily comprehensible pictorial representation, but also bypasses the need to perform detailed elastic–plastic calculations and permits the inclusion of the effects of

CRACK TIP PLASTICITY AND ASSOCIATED EFFECTS

thermal and residual stresses. Moreover, experimental data indicates that the FAD is a lower bound, and hence safe. Further details are contained in [21].

References

1. Spencer, G. S. (1968), *An Introduction to Plasticity*, Chapman and Hall, London.
2. Johnson, W. and Mellor, P. B. (1962), *Plasticity for Mechanical Engineers*, Van Nostrand, New York.
3. Irwin, G. R. (1958), *Fracture, Handbuch der Physik, VI*, pp. 551–90, Springer-Verlag, Heidelberg.
4. Dugdale, D. S. (1960), 'Yielding of steel sheets containing slits', *J. Mech. Phys. Solids*, 8, 100–8.
5. Barenblatt, G. I. (1962), 'The mathematical theory of equilibrium cracks in brittle fracture', *Adv. Appl. Mech.*, 7, 55–129.
6. Burdekin, F. M. and Stone, D. E. W. (1966), 'The crack opening displacement approach to fracture mechanics in yielding materials', *J. Strain Analysis*, 1, 145–53.
7. Shih, C. F. (1974), *Small Scale Yielding Analysis of Mixed-Mode Plane – Strain Crack Problems*, ASTM, STP 560, 187–210.
8. Hahn, G. T. and Rosenfield, A. R. (1965), 'Local yielding and extension of a crack under plane stress', *Acta Metall.*, 13, 293–306.
9. *Fracture Toughness Evaluation by R-Curve Methods* (1973) ASTM, STP 527.
10. Recommended practice for R-curve determination (1975), *ASTM Annual Book of Standards*, Part 10.
11. Weiss, V. and Sengupta, M. (1976), *Ductility, Fracture Resistance and R-Curves* ASTM, STP 590, 194–207.
12. *Methods of Test for Plane Strain Fracture Toughness of Metallic Materials* (1977), BS 5447, British Standards Institute.
13. *Standard Test Method for Plane-Strain Fracture Toughness of Metallic Materials* (1978), ANSI/ASTM E399, Annual Book of ASTM Standards, Part 10.
14. Wells, A. A. (1961), *'Unstable crack propagation in metals – cleavage and fast fracture'*, Cranfield Crack Propagation Symp., 1, 210–30.
15. Robinson, J. N. and Tetelman, A. S. (1974), 'Measurement of K_{Ic} on small specimens using critical crack tip opening displacement', *Fracture Toughness and Slow Stable Cracking*, ASTM STP 559, 139–58.
16. Knott, J. F. (1973), *Fundamentals of Fracture Mechanics*, Butterworth, Sevenoaks.
17. Broek, D. (1974), *Elementary Engineering Fracture Mechanics*, Noordhoff Int., Leiden.
18. Rolfe, S. T. and Barsom, J. M. (1977), *Fracture and Fatigue Control in Structures*, Prentice Hall, New Jersey.

19. Rice, J. R. (1968), 'A path independent integral and the approximate analysis of strain concentrations by notches and cracks', *Trans. ASME, J. Appl. Mech.*, **35**, 379–86.
20. Paris, P. C., Tada, H., Zahoor, A., Ernst, H. (1979), *The Theory of Instability of the Tearing Mode of Elastic-Plastic Crack Growth*, ASTM, STP 668, 5–36.
21. Chell, G. G. (1979), *A Procedure for Incorporating Thermal and Residual Stresses into the Concept of a Failure Assessment Diagram*, ASTM STP 668, 581–605.

7 Fatigue Crack Growth

7.1 Introduction

We have seen that cracks within a structure can produce catastrophic failure when the applied loading causes the stress intensity at the crack tip to reach a critical value. However, it is also possible to induce failure as a result of subcritical crack growth causing an increase in crack length, and hence in stress intensity. One method by which such crack extension may occur is under remote cyclic loading, with maximum and minimum load levels which would not cause failure if applied monotonically, and maintained.

The traditional approach to design under cyclic loading has involved $S-N$ (Stress range versus number of cycles to final failure) data, obtained from as-manufactured materials, with corrections for notch effects, etc. [1]. This approach does not separate out crack initiation and propagation stages, and cannot be applied to the determination of structural life-expectancy with a crack-like defect of a known size.

By characterizing subcritical crack growth using linear elastic fracture mechanics parameters, it is possible to predict crack growth rates under cyclic loading, and hence the number of cycles required for a crack to extend from some initial length to a predetermined length of interest to the designer.

7.2 Stress intensity factor range

Consider a body containing a crack of length $2a$, subjected to a constant-amplitude cyclic loading such that the remotely-applied loading varies between σ_{max} and σ_{min}, the maximum and minimum values of the stress respectively, (Fig. 7.1). Assume at this stage that $\sigma_{max} > \sigma_{min} > 0$, and that loading and geometry are symmetrical about the crack line.

The opening-mode crack tip stress intensity factor at σ_{max}, K_{max}, is given by equation (3.75):

$$K_{max} = Q\sigma_{max}(\pi a)^{1/2} ; \qquad (7.1)$$

similarly

$$K_{min} = Q\sigma_{min}(\pi a)^{1/2} \qquad (7.2)$$

where Q is the configuration correction factor.

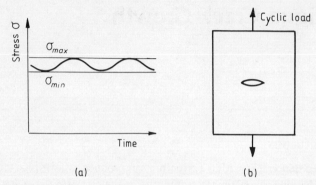

Fig. 7.1 *Fatigue loading:* (a) *Constant-amplitude cyclic loading;* (b) *specimen configuration.*

If we consider the conditions in the immediate vicinity of the crack tip, referring to Section 6.2 we note that there is a plastic zone, whose radius is given in equation (6.6), at the maximum stress level this radius is:

$$\lambda = \frac{1}{2\pi} \left(\frac{K_{max}}{Y} \right)^2.$$

This is the size of the zone which has undergone permanent deformation and the subsequent unloading must accommodate this deformed zone, giving rise to 'reversed' yielding. The process is akin to the nervous to-and-fro bending of a paper clip, which produces its failure. In like manner, the crack extends into the small zone in which reversed yielding is constantly occurring, a new plastic zone is created and the process repeated. In fact, reversed yielding is limited to a zone whose radius is equal to $\lambda/4$, [2]. Since we can characterize the plastic zone dimensions in terms of K, and the changes in these dimensions in terms of changes in K, termed the stress intensity factor range, ΔK:

$$\Delta K = K_{max} - K_{min} = Q\, \Delta\sigma(\pi a)^{1/2} \tag{7.3}$$

where $\Delta\sigma = \sigma_{max} - \sigma_{min}$, it may be advantageous to quantify fatigue crack growth in terms of ΔK also.

7.3 Empirical crack growth rate results

A cracked specimen is subjected to constant-amplitude load cycling, stress range $\Delta\sigma_1$ and increments of crack length are measured, the corresponding number of loading cycles being noted. The resulting data are illustrated schematically in Fig. 7.2, as a plot of semi-crack length, a, versus number of loading cycles, N. (Note that a considerable proportion of the life of the specimen is spent at short crack lengths.)

FATIGUE CRACK GROWTH

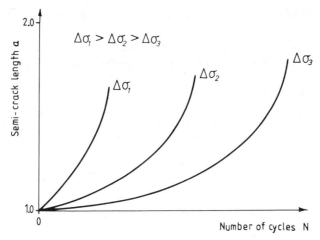

Fig. 7.2 *Typical dependence of crack growth on cyclic stress range.*

By performing experiments on identical specimens subjected to higher stress ranges, $\Delta\sigma_2$ and $\Delta\sigma_3$, two more plots are obtained (Fig. 7.2). Taking each of the three curves in turn, and extracting the slope, da/dN, at a given crack length, and ΔK for the same crack length via equation (7.3), we may plot $\log(da/dN)$ versus $\log(\Delta K)$ (Fig. 7.3).

In fact, since ΔK includes both the effect of crack length changes and the magnitude of loading cycles on the plastic zone dimensions, the three curves of Fig. 7.2 reduce to a single curve when plotted as $\log(da/dN)$ versus $\log(\Delta K)$

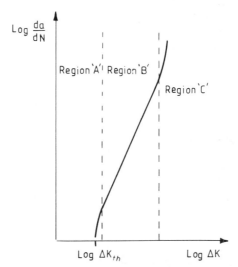

Fig. 7.3 *Schematic representation of fatigue crack growth.*

(Fig. 7.3). The fatigue crack growth characteristics illustrated in Fig. 7.3 fall into three distinct regions. Region A begins with a threshold value of stress intensity, K_{th}, below which crack propagation does not occur, and continues until the slope of the curve becomes constant. For our purposes, the value of K_{th} may be associated with the attainment of a sufficient level of activity in the crack tip region [3]. Region C also exhibits a steep gradient, and is complicated by the possible attainment of plastic zone dimensions which are large compared with specimen dimensions, ductile tearing, and values of K_{max} that approach the fracture toughness, leading to crack extension which includes monotonic components [3, 4].

Region B represents the zone in which the plot is effectively linear, and hence fatigue crack growth can be represented by the relationship:

$$da/dN = C(\Delta K)^m \tag{7.4}$$

where C and m are experimentally determined material constants. The value of m is approximately 3 for steels, and in the range 3–4 for aluminium alloys. The life of most cracked engineering structures may be considered solely in this range, once allowance has been made for the minimum crack length employed, which is normally related to the limitations of a particular inspection technique or design code requirements.

Equation (7.4), is normally called Paris' law [5], and although somewhat modified on occasions, has gained general acceptance. Note that the Paris equation is only an empirical relationship.

7.4 Use of crack growth law

By re-arranging equation (7.4) and integrating we can obtain the number of cycles, ΔN required for a crack to propagate from an initial length, a_i, to some final length, a_f, namely:

$$\Delta N = \int_{a_i}^{a_f} \frac{1}{C(\Delta K)^m} \, da. \tag{7.5}$$

Substituting for ΔK from (7.3) we obtain:

$$\Delta N = \frac{1}{C(\Delta \sigma)^m \pi^{m/2}} \int_{a_i}^{a_f} \frac{1}{Q^m a^{m/2}} \, da. \tag{7.6}$$

Assuming for the moment that Q is independent of a, equation (7.6) may be written:

$$\Delta N = \frac{1}{CQ^m \pi^{m/2}(\Delta \sigma)^m} \int_{a_i}^{a_f} a^{-m/2} \, da. \tag{7.7}$$

FATIGUE CRACK GROWTH

Performing the integration:

$$\Delta N = \frac{1}{CQ^m \pi^{m/2} (\Delta\sigma)^m} \frac{a_i^{1-(m/2)} - a_f^{1-(m/2)}}{m/2 - 1} \tag{7.8}$$

Example 7.1

A large aluminium alloy plate contains a crack of length 1 cm emanating from a circular cutout as shown in Fig. 7.4. The plate is subjected to a constant-amplitude tensile cyclic loading from 6 MN m^{-2} to 60 MN m^{-2}. The Paris law exponent is 3, and ΔK at $da/dN = 10^{-9}$ m/cycle is 2.8 MN m$^{-3/2}$. Assuming that Q is adequately represented by a constant value of 1.02, how many loading cycles must be applied for the crack to grow to 2 cm?

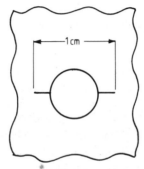

Fig. 7.4 Cracked, circular cutout (Example 7.1).

Solution: First, evaluate the constant 'C' in Paris' law, from equation (7.4):

$$C = \frac{10^{-9}}{2.8^3} = 4.55 \times 10^{-11}, \text{ for crack growth in m/cycle.}$$

Obtain stress range, $\Delta\sigma$:

$$\Delta\sigma = (60 - 6) = 54 \text{ MN m}^{-2}.$$

Now substitute values into equation (7.8) to obtain ΔN:

$$\Delta N = \frac{1}{4.55 \times 10^{-11} \cdot 1.02^3 \cdot \pi^{3/2} \cdot 54^3} \left(\frac{(0.005)^{-1/2} - (0.01)^{-1/2}}{\frac{1}{2}} \right).$$

Thus:

$$\Delta N = 195\,675 \text{ cycles.}$$

Thus far we have assumed that Q remains constant as the crack extends. In Chapter 3 we noted that, in general, Q is a function of both loading and geometry, and since crack length is a major geometrical factor, it may be necessary to incorporate the dependence of Q on a. In this case we must use equation (7.6), which may be written:

$$\Delta N = \frac{1}{C(\Delta\sigma)^m \pi^{m/2}} \int_{a_i}^{a_f} Q(a)^{-m} a^{-m/2} \, da. \tag{7.9}$$

If stress intensity data have been curve-fitted, they may be presented in polynomial form [6]. In this case:

$$Q = \sum_{n=1}^{M} D_n a^n$$

where D_n's are real coefficients and n, M are integers. Integration is then straightforward, and may be performed exactly.

In cases where Q is presented in more complex form it may be necessary to complete the integration numerically (e.g. using Simpson's rule). Alternatively, in such circumstances, or when results are presented in graphical form, a straightforward approximate procedure outlined in [7] may be followed in order to obtain ΔN-values. The method consists of extracting Q-data from the curve of Q versus crack length, shown schematically in Fig. 7.5(a), for crack lengths a_i and a_f, and as many intermediate lengths as required. In the case of fluctuating loads, the stress intensity range will be obtained by evaluating $Q\Delta\sigma(\pi a)^{1/2}$ at each crack length to obtain stress intensity factor ranges $\Delta K_i, \Delta K_1, \Delta K_f$ which correspond to a_i, a_1, a_f, a_1 being an intermediate crack length. Entering the crack growth rate plot (Fig. 7.5(b)) with these data yields values of $da/dN_i, da/dN_1, da/dN_f$ as shown in Table 7.1.

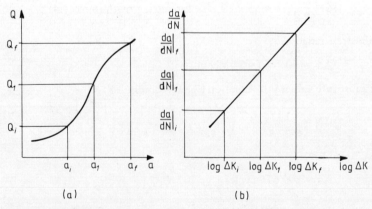

Fig. 7.5 *Graphical determination of crack growth rates:* (a) *Q derivation;* (b) *crack growth rates.*

FATIGUE CRACK GROWTH

Table 7.1

(1) $a(m)$	(2) $da(m)$	(3) $\Delta K \,(\text{MN m}^{-3/2})$	(4) $da/dN \,(\text{m/cycle})$	(5) $da/dN \,(\text{m/cycle})$ geometric mean	(6) dN (cycles) $= da/(da/dN)$
a_i		K_i	da/dN_i		
	$a_1 - a_i$			$\dfrac{da/dN_i + da/dN_1}{2}$	$\dfrac{2(a_1 - a_i)}{da/dN_i + da/dN_1}$
a_1		K_1	da/dN_1		
	$a_f - a_1$			$\dfrac{da/dN_1 + da/dN_f}{2}$	$\dfrac{2(a_f - a_1)}{da/dN_1 + da/dN_f}$
a_f		K_f	da/dN_f		
				Total cycles	

Thus far we have derived the first four columns shown in the table above. By averaging adjacent crack growth rates we may obtain a mean value for each interval, column 5. With this information together with the increment in crack length it is possible to approximate the number of cycles taken in traversing each interval, column 6.

Example 7.2

The stress intensity factor for a crack in an infinite array subjected to remote uniaxial tension σ was derived in Example 3.4, and is:

$$K = \sigma \left[2b \tan\left(\frac{\pi a}{2b}\right) \right]^{1/2} \qquad (7.10)$$

where $2a$ is the crack length, and $2b$ the distance between crack centres. Consider a particular case of a steel plate, $a = 5$ mm, $b = 20$ mm, which is subjected to a constant-amplitude remote loading from zero to 130 MN m^{-2}. If the Paris' Law exponent is 3.3, and $\Delta K = 6.1$ MN m$^{-3/2}$ at $da/dN = 10^{-9}$ m/cycle, calculate the number of cycles for the crack to propagate from 5 mm to 7 mm, and from 10 mm to 12 mm.

Solution: Evaluating the constant 'C' in Paris' law from equation (7.4):

$$C = \frac{10^{-9}}{6.1^{3.3}} = 2.56 \times 10^{-12}, \quad \text{for crack growth in m/cycle.}$$

Constructing a table in accordance with the procedure detailed on the previous page, we note that equation (7.10) yields:

$$K_{5\,\text{mm}} = 130 \left[2(0.02) \tan\left(\frac{\pi(0.005)}{2(0.02)}\right) \right]^{1/2} = 16.73 \text{ MN m}^{-3/2}.$$

Similarly

$$K_{6\,\text{mm}} = 18.56 \text{ MN m}^{-3/2}, \quad K_{7\,\text{mm}} = 20.35 \text{ MN m}^{-3/2}.$$

These values, with the crack growth rates derived from equation (7.4) appear in Table 7.2, together with the averaged crack growth rates, and number of cycles required to propagate over the appropriate ranges.

Notice the increase in crack growth rate by a factor of approximately two, which occurs over the range considered. Now, repeating the process for crack lengths of 10 mm, 11 mm and 12 mm in order to determine number of cycles in the second case we obtain a table of the form shown in Table 7.3.

This indicates the dramatic increase in crack growth, and consequential reduction in life, at longer crack lengths. In fact the number of cycles required to increment the crack by 2 mm has been reduced by a factor of four.

FATIGUE CRACK GROWTH

Table 7.2

a(m)	da(m)	ΔK (MN m$^{-3/2}$)	da/dN (m/cycle)	da/dN (m/cycle) geometric mean	dN (cycles) = da/(da/dN)
0.005		16.73	2.79×10^{-8}		
	0.001			3.36×10^{-8}	29762
0.006		18.56	3.93×10^{-8}		
	0.001			4.63×10^{-8}	21598
0.007		20.35	5.33×10^{-8}		
				Total	51360

Table 7.3

a(m)	da(m)	ΔK (MN m$^{-3/2}$)	da/dN (m/cycle)	da/dN (m/cycle) geometric mean	dN (cycles) = da/(da/dN)
0.010		26.00	1.20×10^{-7}		
	0.001			1.375×10^{-7}	7273
0.011		28.13	1.55×10^{-7}		
	0.001			1.79×10^{-7}	5587
0.012		30.50	2.03×10^{-7}		
				Total	12860

7.5 Other factors affecting fatigue crack growth rate

Fig. 7.6 illustrates two possible forms of constant-amplitude cyclic loading applied to a cracked specimen, both have the same stress intensity factor range, ΔK, since:

$$(K_{\max 1} - K_{\min 1}) = (K_{\max 2} - K_{\min 2}) = \Delta K.$$

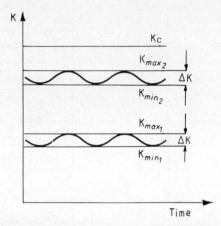

Fig. 7.6 *Two examples of constant-amplitude loading having same value of stress intensity factor range, ΔK.*

However, it may be anticipated that the relative proximity of $K_{\max 2}$ to the fracture toughness, K_c, will affect the crack growth rate. In order to quantify the amount by which the mean stress value during cyclic loading is removed from zero, it is common practice to specify an 'R'-value for a particular loading, where:

$$R = K_{\min}/K_{\max}. \qquad (7.11)$$

Whilst the effect of varying $R (R > 0)$ is generally somewhat limited in the case of steels [4, 8], aluminium alloys may be sensitive to R-value. A particular example, for an aluminium alloy is illustrated schematically in Fig. 7.7, and serves to emphasize the need to select the R-value applicable to the loading.

If the value of $K_{\min}$ within a cycle is negative, this indicates closure at the crack tip at $K = 0$ and as a simple model the corrected ΔK- and R-values may be taken as:

$$\Delta K = K_{\max} \qquad R = 0.$$

In Example 4.3 we saw that it is possible to derive positive stress intensity solutions when crack surfaces are theoretically 'overlapping' one another in a physically unacceptable fashion. When calculating ΔK-values it is advisable to check for this effect, and correct as necessary. This phenomenon may occur

FATIGUE CRACK GROWTH

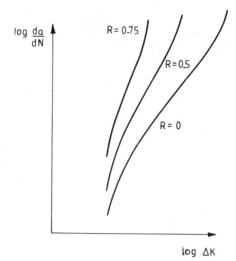

Fig. 7.7 *Effect on crack growth rate of varying R ratio (schematic).*

when superimposing residual stress fields (e.g. those induced by autofrettage or welding) and various remotely applied alternating loads.

7.6 Variable-amplitude loading

Structures are normally subjected to a wide range of load amplitudes. If a single cycle overload is applied during a constant-amplitude fatigue test (Fig. 7.8(a)) the crack growth rate for a period following the overload will be appreciably reduced (Fig. 7.8(b)).

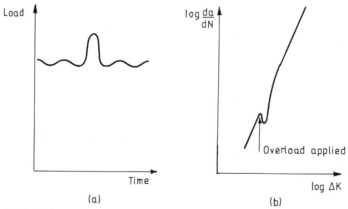

Fig. 7.8 *Variable-amplitude loading:* (a) *single overload;* (b) *effect of overload on crack growth rate.*

The effect of the overload cycle is to induce a relatively large plastic zone ahead of the crack tip, which in turn produces compressive stresses of yield stress magnitude ahead of the crack, into which growth must proceed for a number of cycles before the crack 'breaks free' of the compressive field. During the period of retardation the crack tip stress intensity has been reduced from the calculated value by the presence of compressive stresses.

In cases where the interaction effects between cycles of differing magnitudes are considered negligible, constant amplitude data may be employed, and integrated on (say) a cycle-by-cycle basis in order to predict crack growth. This will produce a conservative estimate since the ignored interaction effects can only serve to retard the crack growth. In general, interaction effects cannot be neglected without incurring excessively conservative life estimates. Some semi-empirical methods aimed at modelling retardation are described in [9], whilst two recent techniques show some promise [10, 11].

7.7 The rainflow method

In some cases individual load cycles obtained from (say) a transducer on an in-service component are difficult to discern (Fig. 7.9(a)). Various methods are available for quantifying the effects of individual cycles. The technique known as 'rainflow counting', or the 'pagoda roof method', has gained some acceptance [12]. Consider the variable loading illustrated schematically in Fig. 7.9(a) to be

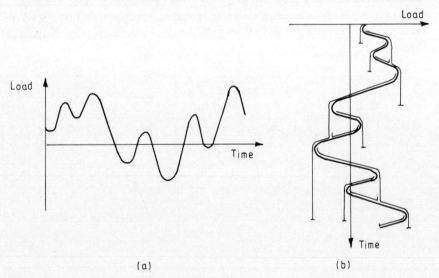

Fig. 7.9 (a) *Schematic load–time record;* (b) *'Rainflow' cycle-counting method applied to the record.*

rotated so that the time axis is vertical, and mentally pour water onto the loading history (Fig. 7.9(b)).

The flow starts at the top (beginning) of the history, and is also initiated at the inside of each maximum or minimum. Flow is stopped when it strikes flow descending from above, or a point opposite a maximum or minimum whose magnitude exceeds that at the point from which it started. Flow also stops at the end of the record. Each flow is a half cycle, and there is a complementary half cycle of opposite sign elsewhere in the complete record, except perhaps for a flow at the beginning or end of the record.

After extracting individual cycles it is a straightforward procedure to sum their individual effects on crack growth rate. The technique does, however, have limitations when complementary half cycles are widely separated in time, so that their contributions to crack growth rate, arising as they do at different crack lengths, are difficult to quantify.

7.8 Mixed-mode loading

Experimental studies of fatigue crack growth under combined mode I and II loading were first conducted by Iida and Kobayashi [13] using a 7075-T6 aluminium alloy with central crack inclined to the tensile axis. The results show that the crack propagation rate is significantly increased by the initial presence of the K_{II} component.

Roberts and Kibler [14] performed experimental work with constant K_I and cyclic K_{II}, and proposed a fatigue crack growth equation which reduces to:

$$\left[\frac{da}{dN}\right]_{II}\bigg|_{R_{II}} = \left[\frac{da}{dN}\right]_{I}\bigg|_{R_I} \left(\frac{K_{Ic}}{K_{IIc}}\right)^{n^*} \tag{7.12}$$

where

$$R_j = \frac{K_{j\,\text{min}}}{K_{j\,\text{max}}}, \quad j = I, II$$

and the exponent n^* differs from the Paris' law exponent. Equation (7.12) is based on assumptions:

(i) That the log (da/dN) versus log (ΔK) curves have the same slope for both modes at all da/dN
(ii) That the mode II curve is shifted by an amount ($K_{Ic} - K_{IIc}$).

More recently Tanaka [15] performed a series of fatigue experiments on inclined cracks in pure aluminium. Tanaka's specimens resemble those of [13] (Fig. 7.10) with inclinations of 90°, 72°, 45°, 30°. Crack length was measured along the growth direction, and much higher growth rates were reported when mode I was accompanied by mode II. Tanaka also analysed, on the basis of models due to

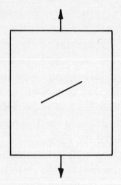

Fig. 7.10 *Specimen containing angled crack.*

Weertman [16] and Lardner [17], the K_I, K_{II} interaction, and proposed the law

$$\frac{da}{dN} = C(K_e)^m \tag{7.13}$$

which is the Paris law (equation (7.4)) with an effective stress intensity factor, K_e. The exponent m is 4 in the Weertman [16] theory, but may increase when the crack tip yielding departs from small scale [17], or when non-striation crack growth mechanisms occur [18]. The best form of K_e in equation (7.13) is given as:

$$K_e = (K_I^4 + 8K_{II}^4)^{1/4}. \tag{7.14}$$

7.9 References

1. Benham, P. P. and Warnock, F. V. (1973), *Mechanics of Solids and Structures*, Pitman, London.
2. Rice, J. R. (1976), 'Mechanics of crack tip deformation and extension by fatigue' in *Fatigue Crack Propagation*, ASTM STP 415.
3. Knott, J. F. (1973), *Fundamentals of Fracture Mechanics*, Butterworth, Sevenoaks.
4. Rolfe, S. T. and Barsom, J. M. (1977), *Fracture and Fatigue Control in Structures*, Prentice Hall, New Jersey.
5. Paris, P. C. and Erdogan, F. (1963), 'A critical analysis of crack propagation laws', *Trans. ASME, J. Bas. Engng*, **85**, 528–34.
6. Rooke, D. P. and Cartwright, D. J. (1976), *Compendium of Stress Intensity Factors*, HMSO, London.
7. *Examples of the Use of Data Items on Fatigue Crack Propagation Rates* (1977), Engineering Sciences Data Unit, Item No. 74017.
8. Pook, L. P. (1975), 'Analysis and application of fatigue crack growth data', *J. Strain Analysis*, **10**, No. 4, 242–50.

9. Broek, D. (1974), *Elementary Engineering Fracture Mechanics*, Noordhoff, Leiden.
10. Gemma, A. E. (1977), 'A new approach to estimate fatigue crack delay due to a single cycle overload', *Engng Frac. Mech.*, **9**, 647–54.
11. Gray, T. D. and Gallagher, J. P. (1976), 'Predicting fatigue crack retardation following a single overload using a modified Wheeler model' in *Mechanics of Crack Growth*, ASTM STP 590.
12. Socie, D. F. (1975), *Fatigue Life Prediction Using Local Stress–Strain Concept*, presented at SESA Spring Meeting, Chicago, Illinois, 11–16 May.
13. Iida, S. and Kobayashi, A. S. (1969), 'Crack propagation rate in 7075–T6 plates under cyclic tensile and transverse shear loadings', *Trans. ASME, J. Bas. Engng*, **91**, 764–769.
14. Roberts, R. and Kibler, J. J. (1971), 'Some aspects of fatigue crack propagation', *Engng Frac. Mech.*, **2**, 243–60.
15. Tanaka, K. (1974), 'Fatigue crack propagation from a crack inclined to the cyclic tensile axis', *Engng Frac. Mech.*, **6**, 493–507.
16. Weertman, J. (1966), 'Rate of growth of fatigue cracks calculated from the theory of infinitesimal dislocations distributed on a plane', *Int. J. Frac. Mech.*, **2**, 460–67.
17. Lardner, R. W. (1968), 'A dislocation model for fatigue crack growth in metals', *Phil. Mag.*, **17**, 71–82.
18. Ritchie, R. O. and Knott, J. F. (1973), '*Brittle cracking processes during fatigue crack propagation*', Proc. 3rd Int. Congr. on Fracture, Munich, V-434/A.

8 The Fracture Mechanics Design Process

8.1 Introduction

In this final chapter we examine the role of the designer in applying fracture mechanics criteria in order to obtain a high level of certainty of structural reliability. Other specialized and extensive publications cover the detailed design techniques in particular branches of engineering [1–6]. There are, of course, other possible modes of failure, e.g. general yielding, buckling, excessive deformation, corrosion and creep. It is implicitly assumed that other modes of failure are considered in parallel with the fracture mechanics design process. We shall firstly attempt to identify the common features and problems in design using fracture mechanics criteria, e.g. flaw detection and sizing, the meaning of 'safety', the search for data on materials and stressing, before considering some straightforward design problems from various branches of engineering.

8.2 Crack detection techniques

There are a range of crack detection techniques available which enable a component to be examined without further damage [7]. The actual size of defect which is likely to be detected varies widely, even for a particular detection technique, and depends on operator experience, crack location, orientation, etc. An important prerequisite for non-destructive testing (NDT) examination is normally a knowledge of critical locations likely to develop cracks during service. Some of the NDT methods available are listed below:

(a) *Visual inspection*. This depends on the use of the naked eye assisted only by visual magnifiers, mirrors and suitable lighting. Considerable user experience is required to detect even moderately short surface cracks. Accessibility is a particular problem which may be mitigated to some extent by the use of fibre-optical methods.
(b) *Dye-penetrant*. Suitable for surface cracks only, the method involves the application of a liquid penetrant, which is subsequently wiped off the surface before application of a powdered 'developer'. Crack-like defects will produce a contrasting coloured line on the developer.
(c) *Thermal*. Relatively little used, depends upon the changes in thermal

THE FRACTURE MECHANICS DESIGN PROCESS

surface pattern (as indicated by chemicals with particular melting point properties, or by infrared detectors) arising from internal or surface defects.

(d) *Magnetic anomalies.* Flourescent liquid containing iron particles in suspension is applied to component. When placed in a strong magnetic field and illuminated with ultraviolet light, disturbances in magnetic field induced by cracks and cutouts appear as a change in the field pattern. Limited to magnetic materials.

(e) *Radiography.* In the case of X-ray or gamma-ray radiography, a portable source is used to irradiate the component, and the absorption assessed from the image on a sensitive film on the opposite side of the component from the source. Cracks, which absorb less radiation, appear as dark areas on the film. The method is sensitive, and may be used to detect internal cracks. However, poor crack orientation may produce inferior images.

(f) *Ultrasonics.* A probe emits a high frequency sound wave into the component, which is reflected by surfaces, including internal cracks. The time taken for transmission and reflection of a pulse is normally indicated on an oscilloscope. This may be interpreted as a distance through the component, and hence allow the crack to be properly located.

(g) *Eddy current.* A small coil induces eddy currents in the metal component. This re-induces a current in the coil. A change in the inductive 'fingerprint' of a component may indicate a crack or defect.

(h) *Acoustic emission.* Stress or pressure waves are generated within components during dynamic material processes. The presence of cracks alters the load at which plastic deformation begins, thus altering the acoustic emission pattern. The record obtained during loading can be used to assess the incidence and severity of flaws [8].

8.3 Initial flaw sizes

Total component lifetime will depend on the time required for crack initiation and propagation from some subcritical initial length to a critical length.

Application of fracture mechanics criteria implies the existence of crack-like defects within the structure. The initial size of such defects may be fixed in one of several different ways:

(a) The actual defect size may be known as a result of an inspection.

(b) On the basis of the known limitations of a particular crack detection technique, the dimensions of a crack which was 'just missed' may be assumed.

(c) The dimensions and location of cracks which must be assumed to exist may be fixed by law, in the form of requirements imposed by a particular code of practice. (Normally termed a 'damage tolerance' requirement.)

8.4 Fail-safe and safe-life design concepts

Two basic design philosophies may be employed in the avoidance of brittle failure using fracture mechanics design techniques. The *fail-safe* concept assumes that, in spite of the failure (or incipient failure) of an individual component, the complete structure has sufficient built-in redundancy (or remaining lifetime) to be safe from overall catastrophic failure, and to permit subsequent repairs to be effected. Fail-safe design may be achieved in several ways:

(a) *Multiple load paths*, such as stringers, may be provided so that failure of one member leads to a redistribution to other, intact members.

(b) *Crack arresters* may be positioned so as to inhibit crack propagation before overall structural failure occurs. This is a familiar technique in aircraft construction, with integral, riveted or bonded stiffeners [6] (Fig. 8.1). In ship-hull design welded in-plane crack arrester plates of higher toughness than the main hull material may be employed (Fig. 8.2) as an alternative to the more familiar welded or riveted out-of-plane arresters [9].

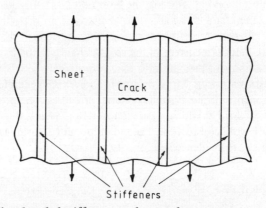

Fig. 8.1 *Riveted or bonded stiffeners used as crack arresters.*

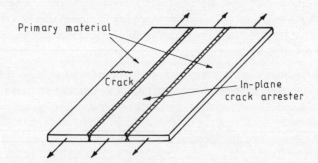

Fig. 8.2 *Welded, in-plane, high-toughness, crack arrester.*

THE FRACTURE MECHANICS DESIGN PROCESS 141

(c) *Inspection intervals* may be selected for fatigue-loaded structures in order to ensure detection of a subcritical fatigue crack before it can reach a critical length. The component may then be replaced, and normal inspection resumed.

The alternative to fail-safe design is the *safe-life* design concept, which is aimed at ensuring that a particular component, or complete structure will not fail in service within its design life. It is usually applied to structural components which experience fatigue loading, whose failure may induce failure of the structure. Thus the designer must know the loading conditions during the life of the structure, and the effect of such loading on the material. This enables the designer to predict service life in terms of number of cycles to failure, or of time to failure. The philosophy normally requires the use of large factors of safety. Large fixed structures, e.g. bridges and buildings, are frequently the subject of safe-life design, either in whole or in part.

8.5 Locating stress intensity factor solutions

For a given fracture mechanics design problem it will be necessary to calculate the configuration correction factor, Q (see Section 3.9), for the geometry and loading under consideration. The immediate ready-made sources of reference available to the designer include several compendia of solutions [10–13]. When using such compendia Q is obtained either from a simple graphical presentation, or by evaluating a simple polynomial or analytic expression with given coefficients.

However, even solutions obtained from such standard sources should be viewed critically in at least two respects. Firstly, does the model selected really fulfil the requirements of the designer's geometry and boundary conditions? Secondly, what is the accuracy associated with the solution? In general it is somewhat difficult to place error bounds on numerical stress intensity factor solutions, but it may be possible to produce reasonably accurate error estimates on the basis of certain convergence criteria, or comparison with other solutions. For instance, a much-used solution is that of a single crack emanating from a circular boundary, under remote loading. An early solution to this problem, still widely used, is due to Bowie [14], who used a mapping–collocation technique. Subsequent comparisons [15, 16] indicate a disagreement of 6% for certain crack lengths. An error of this magnitude will be magnified to one of order 20% in terms of crack growth rate, assuming an exponent of 3 in the Paris crack growth equation (Section 7.3).

If the solution sought is not immediately available, there may be an appropriate Green's function (Section 4.3) or weight function (Section 4.4) for the problem geometry, from which an accurate solution may be determined with relatively little effort. For example, a cracked, welded component having complex residual stresses, which may be modelled as an edge-crack or a centrally located internal crack in a finite width strip may be solved in a straightforward manner using published Green's functions of high accuracy. Next the designer may choose to conduct a literature search to locate the required solution, or contact appropriate agencies or individuals directly. The only problem is in selecting the possible sources for solution. A little thought will indicate that cracked-stiffened structures will be of concern to the aeronautical industry and associated research establishments; complex residual stress, and thermal stress problems arise in the welding and nuclear industries; whilst a cracked, pin-loaded lug may be a problem for a bridge designer but may already have been solved for the purposes of undercarriage design.

If an accurate solution is not immediately available, there are a range of approximate methods [17], e.g. compounding (Section 4.9), weight functions (Section 4.4), available to the designer, which may produce rapid results. In the event that the problem defies all of the foregoing options, the designer must ask whether the effort of an 'in-house' analysis is worthwhile. If so, he may elect to use finite element methods (Section 4.6), most packages now possess a 'cracked element' facility, or perhaps a relatively sophisticated modified mapping—collocation (MMC) technique (Section 4.5.5). An MMC package to cope with complex two-dimensional problems is now available [18].

8.6 Locating fracture toughness and fatigue crack growth data

From the frequency with which fracture mechanics and fatigue strength and life predictions are now made the designer may expect that standard reviews and works of reference should be readily available. Apart from short listings [19], most reviews of fracture toughness data have been confined to metallic materials within a particular field of engineering [20–22]. However, a recent compendium includes an attempt at a listing of data sources for all metallic alloys [23].

Various data sheets containing crack growth data are readily available [24, 25]. Once again, a full review of relevant sources is included in [23].

It is worthwhile to caution that the data should be selected critically, and used with care. We have considered some of the factors which can affect toughness and fatigue crack growth rate, e.g. material thickness, test temperature, environment, loading range. The designer must check that all factors are adequately covered.

In the event that appropriate data cannot be located, there is still the option of asking appropriate authorities, or of performing the required tests 'in-house'.

8.7 Design examples

We shall now cover a range of specific design examples in order to apply some of the techniques described in the preceding pages. The examples are graded, and

THE FRACTURE MECHANICS DESIGN PROCESS

the reader may find it worthwhile to work through all of the examples. Example 8.1 is a straightforward life-assessment, intended to illustrate a general approach, Example 8.2 refers to a welded component wherein the configuration correction depends solely upon the loading, however a similar approach will apply to any problem involving residual stress fields. Example 8.3 applies to the design of a rotating component with significant geometrical and loading effects.

The final examples are somewhat more involved. Example 8.4 covers a complex geometry and loading, with a superimposed residual stress field aimed at enhancing component lifetime, whilst Example 8.5 requires intuition in the derivation of a stress intensity factor.

Several of the examples involve iterative procedures which are well suited to numerical computation. This approach has not been emphasized, but readers with available equipment and expertise may wish to apply numerical techniques.

Example 8.1

A long steel plate has a thickness of 3 cm and a width of 30 cm. On the basis of the inspection technique employed, an initial edge crack of length 8.5 mm is assumed to exist, (Fig. 8.3). The plate is subjected to a remotely applied tensile force which varies cyclically between 1.8 MN and 2.7 MN. A lifetime of 90 000 cycles is required from the component. Does it meet this requirement? Discuss critically the options open to the designer in seeking an improved lifetime.

(Material properties: K_{Ic} = 80 MN m$^{-3/2}$, ΔK at 10^{-9} m/cycle = 5.1 MN m$^{-3/2}$, Paris law exponent m = 3.3.)

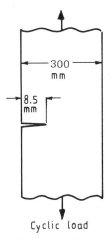

Fig. 8.3 *Edge-cracked plate configuration (Example 8.1)*.

Solution:

Cross-sectional area of specimen = 9000 mm^2

Maximum cyclic stress, σ_{max} = 2.7/0.009 = 300 MN m^{-2}

Minimum cyclic stress, σ_{min} = 1.8/0.009 = 200 MN m^{-2}

Referring to *Compendium of Stress Intensity Factors* [10], Section 1.1.20, we obtain a solution for the edge-cracked strip in tension. This appears both in graphical form, and as a polynomial given as [26]:

$$Q = K_I/K_0 = 1.12 - 0.23\,(a/b) + 10.6\,(a/b)^2 - 21.7\,(a/b)^3 + 30.4\,(a/b)^4 \qquad (8.1)$$

where a is the crack length, b is the plate width and $K_0 = \sigma(\pi a)^{1/2}$. The critical value a_{cr} of crack length, is obtained from:

$$a_{cr} = \left(\frac{K_{Ic}}{Q\pi^{1/2}\sigma_{max}}\right)^2.$$

Assume as a first estimate that Q = 1.12, thus we obtain:

$$a_{cr} = \left(\frac{80}{1.12\pi^{1/2}300}\right)^2 = 0.01804 \text{ m}$$

with this crack length, a/b becomes 0.06. Substitution into (8.1) gives Q = 1.14. This modifies our expression for a_{cr} to give:

$$a_{cr} = \left(\frac{80}{1.14\pi^{1/2}300}\right)^2 = 0.01742 \text{ m}.$$

One more iteration gives a Q-value of 1.139, and a_{cr} = 0.01746 m. In the case considered, the variation in Q is small for the whole range of crack lengths up to a_{cr}, and we shall assume a constant value, Q = 1.14, in order to predict life-expectancy. Equation (7.7) applies to the case of a constant Q-value, and gives the number of cycles, ΔN, to propagate from initial length, a_i, to final length, a_f:

$$\Delta N = \frac{1}{CQ^m \pi^{m/2}(\Delta\sigma)^m}\left(\frac{a_i^{1-(m/2)} - a_f^{1-(m/2)}}{m/2 - 1}\right).$$

From Paris' crack growth rate equation (7.4), $da/dN = C(\Delta K)^m$. Solving for C from the information given:

$$C = 4.624 \times 10^{-12}, \quad \text{for crack growth in m/cycle.}$$

THE FRACTURE MECHANICS DESIGN PROCESS

Now, taking care to ensure that the units are consistent, we obtain:

$$\Delta N = \frac{1}{(4.624 \times 10^{-12})(1.14)^{3.3}\pi^{1.65}100^{3.3}} \times \left(\frac{0.0085^{-0.65} - 0.01746^{-0.65}}{0.65}\right) = 68\,036 \text{ cycles.}$$

Thus the predicted component lifetime does not meet the requirement of 90 000 cycles, even in the absence of a safety factor.

The options available to the designer for life improvement are:

(a) Employing a component with a higher K_{Ic} so that the critical crack size and hence lifetime to failure, will be increased. (This assumes that fatigue crack growth characteristics are unchanged.)
(b) Reducing the maximum value of the applied load, and hence σ_{max}, causing the critical crack size to be increased, and lifetime improved.
(c) Reducing $\Delta\sigma$ to reduce the rate of crack growth.
(d) Improving the inspection technique, and hence reducing the assumed size of the initial defect. For example, by reducing a_i from 8.5 mm to 6 mm we may recalculate the life and obtain 114 280 cycles to failure. In this case the component lifetime is increased above the required 90 000 cycles.

Example 8.2

As a result of welding processes, components may contain residual stresses of yield stress magnitude after fabrication. One such component which has not been stress-relieved contains a residual stress field shown, for the region of interest, in Fig. 8.4.

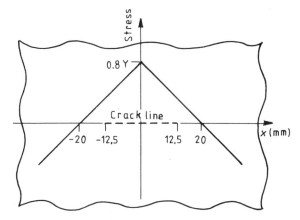

Fig. 8.4 *Unflawed residual stress field (Example 8.2).*

A through-crack of total length 25 mm is detected, and is also shown in Fig. 8.4.

(a) The component is to be subjected to a remote tensile stress. At what stress level will the component fail? To what extent would this be improved by stress-relief prior to loading?
(b) If the component with residual stresses is subjected to cyclic compressive loading, $\sigma_{min} = -300$ MN m^{-2}, $\sigma_{max} = -100$ MN m^{-2}, how many loading cycles may be applied before total crack length reaches 50 mm? Will the crack arrest?

(Material properties: $K_{Ic} = 120$ MN m$^{-3/2}$, ΔK at 10^{-9} m/cycle = 6.1 MN m$^{-3/2}$, Paris' law exponent $m = 3.3$, material yield stress $Y = 1000$ MN m^{-2}.) For the purposes of this example assume that the plate is infinitely large, and that configuration corrections arise solely from the form of the loading.

Solution: (a) Stress intensity factors for a through crack of length $2a$ in an infinite plate may be derived via the weight function expression (equation (4.21)) as:

$$K = 2\left(\frac{a}{\pi}\right)^{1/2} \int_0^a \frac{p(x)}{(a^2 - x^2)^{1/2}} dx$$

For a constant crack-line pressure, p, we obtain the stress intensity factor:

$$K = p(\pi a)^{1/2} \tag{8.2}$$

whilst for a loading which varies linearly with $|x|$, $p(x) = B|x|$:

$$K = 2pBa^{3/2}\pi^{-1/2}. \tag{8.3}$$

Thus, for the case of a crack of half length 12.5 mm, the residual stress over the crack line, σ_r, may be written:

$$\sigma_r = 0.8Y[1 - 0.625(x/a)] \tag{8.4}$$

since, in terms of equations (8.2) and (8.3), $p = 0.8Y$ and $B = (-0.625/a)$. Substituting into (4.21) to obtain the stress intensity factor due to residual stresses, K_r:

$$K_r = 0.8Y(\pi a)^{1/2}[1 - (1.25/\pi)].$$

Substituting for yield stress $Y = 1000$ MN m^{-2}, and crack length $a = 0.0125$ m:

$$K_r = 95.33 \text{ MN m}^{-3/2}.$$

At failure:

$$K_{Ic} = K_r + K_\sigma$$

where K_σ is the stress intensity due to remotely applied stress σ. Thus, at failure:

$$K_\sigma = 120 - 95.33 = 24.67 \text{ MN m}^{-3/2}.$$

THE FRACTURE MECHANICS DESIGN PROCESS   147

But $K_\sigma = \sigma(\pi a)^{1/2}$, thus the stress level at failure is given by:

$$\sigma = \frac{24.67}{(0.0125\pi)^{1/2}} = 124.5 \text{ MN m}^{-2}.$$

If the component is stress-relieved prior to use, $K_r = 0$ and the stress level at failure becomes:

$$\sigma = \frac{120}{(0.0125\pi)^{1/2}} = 605 \text{ MN m}^{-2}.$$

(b) Following the procedure outlined in Example 7.2, Section 7.4, we construct Table 8.1 to enable us to calculate crack growth rates, and hence total life. Table 8.1 will be constructed for increments in crack length of 2.5 mm, from 12.5 mm to 25 mm. For $a = 12.5$ mm:

$$K_{\sigma max} = -100(0.0125\pi)^{1/2} = -19.68 \text{ MN m}^{-3/2}$$

$$K_{\sigma min} = -300(0.0125\pi)^{1/2} = -59.45 \text{ MN m}^{-3/2}$$

whilst the residual stress contribution has already been obtained as:

$$K_r = +95.33 \text{ MN m}^{-3/2}.$$

The total stress intensities at the extremes of the loading cycle are:

$$K_{max} = K_{\sigma max} + K_r = 75.65 \text{ MN m}^{-3/2}$$

$$K_{min} = K_{\sigma min} + K_r = 35.88 \text{ MN m}^{-3/2}.$$

Finally, the stress intensity factor range is simply:

$$\Delta K = K_{max} - K_{min} = 39.77 \text{ MN m}^{-3/2}.$$

This value is shown in Table 8.1.

Clearly, ΔK is exactly the value of stress intensity range which is obtained if the K_r contribution is ignored. This will only affect crack growth rate if we are intending to correct for the R ratio. (Recall that the R ratio defined in Section 7.5 is given by $R = K_{min}/K_{max}$). In the case considered, neglecting K_r produces an R-value given by:

$$R = K_{\sigma min}/K_{\sigma max} = 3.02$$

whilst taking account of K_r we obtain:

$$R = K_{min}/K_{max} = 0.474.$$

However, we shall neglect the effect of changing R ratio, and note a far more dramatic effect arising from the presence of residual stresses. Consider the case

Table 8.1 Crack growth in welded component

a(m)	da(m)	ΔK(MN m$^{-3/2}$)	da/dN(m/cycle)	da/dN(m/cycle) geometric mean	dN(cycles) = da/(da/dN)
0.0125					
	0.0025	39.77	4.809 × 10^{-7}	5.648 × 10^{-7}	4427
0.015					
	0.0025	43.40	6.486 × 10^{-7}	7.432 × 10^{-7}	3364
0.0175					
	0.0025	46.90	8.377 × 10^{-7}	8.650 × 10^{-7}	2890
0.020					
	0.0025	47.80	8.922 × 10^{-7}	5.878 × 10^{-7}	4253
0.0225					
	0.0025	33.77	2.834 × 10^{-7}	1.587 × 10^{-7}	15754
0.025					
		17.76	0.340 × 10^{-7}		
				Total	30 688

THE FRACTURE MECHANICS DESIGN PROCESS 149

$a = 22.5$ mm:

$$K_{\sigma_{max}} = -100\,(0.0225\pi)^{1/2} = -26.59 \text{ MN m}^{-3/2}$$
$$K_{\sigma_{min}} = -300\,(0.0225\pi)^{1/2} = -79.76 \text{ MN m}^{-3/2}$$

The residual stress contribution for this case must be calculated using:

$$\sigma_r = 0.8Y[1 - 1.125(x/a)]$$

which gives:

$$K_r = 0.8Y(\pi a)^{1/2}[1 - (2.25/\pi)] = +60.36 \text{ MN m}^{-3/2}.$$

The total stress intensities at the extremes of the loading cycle are:

$$K_{max} = K_{\sigma_{max}} + K_r = 33.77 \text{ MN m}^{-3/2}$$
$$K_{min} = K_{\sigma_{min}} + K_r = -19.4 \text{ MN m}^{-3/2}.$$

A negative opening-mode stress intensity factor implies physically unacceptable overlapping of the crack surfaces (Section 7.5). As crack closure occurs during the cycle, the minimum value of stress intensity is zero. Thus:

$$\Delta K = K_{max} - 0 = 33.77 \text{ MN m}^{-3/2}.$$

Obviously, in this case it would be incorrect to ignore the contribution from K_r.

The full set of ΔK-values, taking account of K_r effects, are given in Table 8.1. In order to calculate crack growth rates we use Paris' law:

$$da/dN = C(\Delta K)^m$$

where $m = 3.3$.

Solving for C from the information given we obtain $C = 2.56 \times 10^{-12}$, for crack growth in m/cycle. Substitution of appropriate ΔK-values yields the instantaneous (da/dN)-values given in Table 8.1. The tabular calculation is completed along the lines of Example 7.2, giving the number of cycles required for the crack to propagate to a total length of 50 mm as 30 688.

Inspection of the ΔK-values indicates that the range of stress intensity is reducing as the crack extends beyond 20 mm, and hence the rate of crack growth is also reduced. Indeed, performing the appropriate calculations for $a = 27.5$ mm we obtain $K_{max} = -0.083$ MN m$^{-3/2}$, $K_{min} = -58.87$ MN m$^{-3/2}$, indicating that the crack is closed for the complete loading cycle, and hence no crack growth will occur.

Example 8.3

A circular rotor, of radius 300 mm rotates at a speed of 10 000 revolutions per minute. The damage tolerance specification requires the designer to assume the existence of a radial edge-crack of length 10 mm (Fig. 8.5).

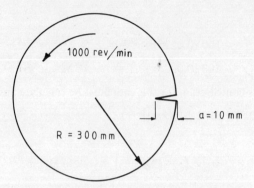

Fig. 8.5 *Cracked rotor configuration (Example 8.3).*

What is the critical crack length? How many times may the rotor be run up to speed before the crack attains a length of 30 mm?

(Material properties: Poisson's ratio, $v = 0.3$, density, $\rho = 7.9 \times 10^3$ kg m^{-3}, $K_{Ic} = 60$ MN m$^{-3/2}$. Paris' law exponent, $m = 3.0$ and $C = 5 \times 10^{-11}$, for crack growth in m/cycle.)

Solution: Q-values for the case of an edge-crack of length a, in a disc which rotates with angular velocity ω are given in a compendium [10]. The graphical data are based on a solution obtained using integral transforms, and of high accuracy [27]. The results are shown graphically, for a Poisson's ratio of 0.3, in Fig. 8.6.

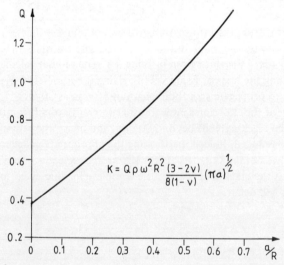

Fig. 8.6 *Configuration correction factor for cracked rotor (Example 8.3) (after [27]).*

THE FRACTURE MECHANICS DESIGN PROCESS 151

The actual stress intensity for plane strain conditions is obtained from the graphical Q-value via the expression:

$$K = Q\rho \frac{\omega^2 R^2}{8} \frac{3-2v}{1-v} (\pi a)^{1/2}. \tag{8.5}$$

In order to determine the crack length at failure, a_{cr}, it will be necessary to perform a simple iterative process. Initially assume $Q = 0.5$ (say) at failure, thus:

$$K_{Ic} = 78 \times 10^6 = (0.5)(7.9 \times 10^3)(1047)^2 \frac{(0.3)^2}{8} \frac{2.4}{0.7} (\pi a)^{1/2}$$

where 1047 rad s^{-1} is equivalent to 10 000 revolutions/minute.
Hence:

$$a_{cr} = 0.0553 \text{ m} = 55.3 \text{ mm}.$$

This is equivalent to $a/R = 55.3/300 = 0.184$. This yields a Q-value of 0.6. Repeating the above calculation with $Q = 0.6$ we obtain $a_{cr} = 48.2$ mm. This process converges to a value $a_{cr} = 37.98$ mm, at a consistent Q-value.

The calculation of lifetime follows the normal tabular routine. The table is constructed for increments in crack length of 5 mm, between 10 and 30 mm. At each crack length we calculate a/R, in order to extract the appropriate Q-value. The stress intensity range is then given by the value of K derived via equation (8.5), since the minimum value of stress intensity occurs when the rotor is at rest, i.e. $K = 0$. The complete listing is included in the Table 8.2.

Crack growth rates are calculated directly from Paris' equation:

$$\frac{da}{dN} = C(\Delta K)^m$$

where C and m are given as 5×10^{-11} and 3.0 respectively, for crack growth in m/cycle. After calculating average values and number of cycles we obtain a total of 9153 cycles for the crack to attain a length of 30 mm.

Example 8.4

The cross section of a prototype gun barrel is shown in Fig. 8.7. Under repeated firing the barrel rapidly develops an array of 40 radial cracks, which undergo fatigue growth as a result of the cyclic pressurization. It is a requirement that the fatigue life of the barrel, based on an initial crack length of 5 mm and pressure of 400 MN m^{-2}, shall exceed the predicted wear life of 10 000 rounds. Does the proposed barrel meet the specification?

As a modification it is proposed that the barrel be fully autofrettaged (a process in which high pressures are applied to the bore during manufacture in order to induce internal yielding, and hence an advantageous compressive stress field near the bore). What is the fatigue life of the modified barrel?

Table 8.2 Crack growth in rotor

a(m)	a/R	Q	da(m)	ΔK(MN m$^{-3/2}$)	$\dfrac{da}{dN}$ (m/cycle)	$\dfrac{da}{dN}$ (m/cycle) geometric mean	dN(cycles) $= da/(da/dN)$
0.01	0.033˙	0.4		23.68	6.641×10^{-7}		
0.015	0.050	0.043	0.005	31.18	1.516×10^{-6}	1.090×10^{-6}	4587
0.02	0.66˙	0.45	0.005	37.68	2.765×10^{-6}	2.141×10^{-6}	2335
0.025	0.083˙	0.475	0.005	44.67	4.457×10^{-6}	3.611×10^{-6}	1385
0.03	0.100	0.515	0.005	52.81	7.364×10^{-6}	5.911×10^{-6}	846
						Total	9153

THE FRACTURE MECHANICS DESIGN PROCESS

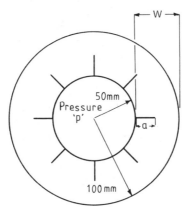

Fig. 8.7 *Gun barrel with multiple radial cracks (Example 8.4).*

A new round is proposed which would operate at a higher pressure. What is the maximum working pressure of the autofrettaged barrel which would meet the original life specification of 10 000 rounds?

(Material properties: Yield stress $Y = 1200$ MN m^{-2}, $K_{Ic} = 90$ MN m$^{-3/2}$, Paris' law $m = 3.1$, $C = 1.455 \times 10^{-11}$, for crack growth in m/cycle.)

Solution: Reference to compendia of solutions does not produce a suitable solution. A literature search reveals an approximate solution [28] with estimated errors of order 7%. The appropriate results for the case of 40 radial cracks are shown in Fig. 8.8, for both autofrettaged and non-autofrettaged tubes with internal pressure, p.

In the case of the non-autofrettaged barrel, an iterative procedure along the lines of Example 8.3 produces values for the critical crack length, a_{cr}, with an internal pressure of 400 MN m^{-2}:

a_{cr} (non-autofrettaged) = 15.8 mm.

Likewise, recognizing that the ratio $Y/p = 3.0$, we obtain the equivalent value for the autofrettaged tube:

a_{cr} (autofrettaged) = 27.75 mm.

The differences in a_{cr} are noteworthy. In the case of the non-autofrettaged tube the critical crack length is short, and barrel failure may be catastrophic. However, in the autofrettaged barrel the crack length is relatively much greater, and the real-life effect of crack curvature may further reduce stress intensity and produce a leak-before-break condition.

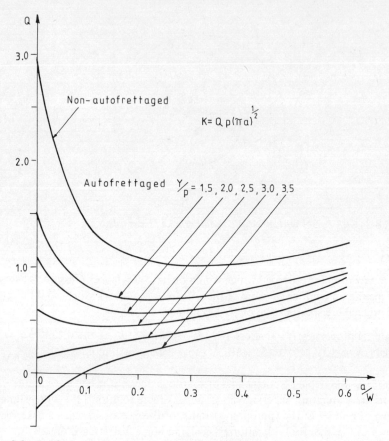

Fig. 8.8 *Configuration correction factor for pressurized thick cylinder with 40 radial cracks (after [28]).*

The reader is by now fully aware of the steps in tabular calculation of fatigue life. Tables 8.3 and 8.4 are constructed for the non-autofrettaged and autofrettaged tubes respectively.

The difference in life-expectancy between the two tubes is dramatic. The results are:

>Non-autofrettaged barrel: 1011 rounds to failure
>Autofrettaged barrel: 45 353 rounds to failure.

Thus, the original barrel fails to meet the life-specification of 10 000 rounds to failure, whilst the modified barrel is perfectly adequate.

Table 8.3 Crack growth in non-autofrettaged barrel (internal pressure 400 MN m^{-2})

a(m)	$\dfrac{a}{W}$	Q	da(m)	ΔK(MN m$^{-3/2}$)	$\dfrac{da}{dN}$ (m/cycle)	$\dfrac{da}{dN}$ (m/cycle) (geometric mean)	dN(cycles) $= da/(da/dN)$
0.005	0.1	1.39		69.68	7.525×10^{-6}		
			0.0025			8.008×10^{-6}	312
0.0075	0.15	1.18		72.45	8.491×10^{-6}		
			0.0025			9.581×10^{-6}	260
0.010	0.2	1.10		77.99	1.067×10^{-5}		
			0.0025			1.177×10^{-5}	212
0.0125	0.25	1.045		82.83	1.286×10^{-5}		
			0.0025			1.411×10^{-5}	177
0.015	0.3	1.10		87.70	1.535×10^{-5}		
			0.008			1.599×10^{-5}	50
0.0158	0.316	1.01		90.00	1.663×10^{-5}		
						Total	1011

Table 8.4 Crack growth in autofrettaged barrel (internal pressure 400 MN m^{-2})

a(m)	$\dfrac{a}{W}$	Q	da(m)	ΔK(MN m$^{-3/2}$)	$\dfrac{da}{dN}$ (m/cycle)	$\dfrac{da}{dN}$ (m/cycle) (geometric mean)	dN(cycles) $= da/(da/dN)$
0.005	0.10	0.25					
0.010	0.20	0.33	0.005	12.53	3.685×10^{-8}	1.462×10^{-7}	34 199
0.015	0.30	0.42	0.005	23.40	2.555×10^{-7}	6.333×10^{-7}	7 895
0.020	0.40	0.54	0.005	36.47	1.011×10^{-6}	2.227×10^{-6}	2 245
0.025	0.50	0.66	0.005	54.14	3.442×10^{-6}	6.253×10^{-6}	800
0.02775	0.555	0.76	0.00275	73.99	9.064×10^{-6}	1.285×10^{-5}	214
				90.00	1.663×10^{-5}	Total	45 353

THE FRACTURE MECHANICS DESIGN PROCESS

Notes

(a) This form of fatigue crack growth under cyclic pressurization may occur in pipelines, (e.g. there is a diurnal variation in pressure in the large-bore sections of natural gas pipelines, as they perform one of their roles as a reservoir for gas).

(b) The technique of life-enhancement arising from judiciously located residual stress fields is the basis of mandrel-enlargement techniques aimed at improving the life of fastener holes, and of shot-peening.

The calculation of maximum cyclic pressure to give a specified life is obviously somewhat complicated, since the pressure affects Y/p ratio, and hence growth rate and critical crack length. By 'guessing' working pressures to bracket a working life of 10 000 rounds it is possible to converge on the desired solution. In the present case the reader may wish to check that a working pressure of 500 MN m^{-2} will produce a fatigue life of 10 017 rounds. The relevant figures are given in Table 8.5, note that for this case $Y/p = 2.5$.

Example 8.5

A pin-loaded lug (Fig. 8.9) forms part of little-used bridge structure in the Arctic zone. It is subjected to a constant tensile load of 250 kN and additional traffic induced loading given by $0.45W$ where W represents the weight of the vehicle in MN. Pin-fretting has induced two diametrically opposed equal length cracks, which are discovered when the tip-to-tip distance has reached 24 mm (Fig. 8.9).

What immediate vehicle weight restrictions should be placed on the use of the bridge?

(Material properties: $K_{Ic} = (0.2T + 70)$ MN m$^{-3/2}$, $-140 \leqslant T \leqslant +150$, where T is the temperature in degrees Centigrade.)

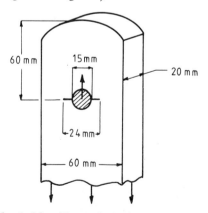

Fig. 8.9 *Cracked pin-loaded lug (Example 8.5).*

Table 8.5 Crack growth in autofrettaged barrel (internal pressure 500 MN m^{-2})

a(m)	$\dfrac{a}{W}$	Q	da(m)	ΔK(MN m$^{-3/2}$)	$\dfrac{da}{dN}$ (m/cycle)	$\dfrac{da}{dN}$ (m/cycle) (geometric mean)	dN(cycles) $= da/(da/dN)$
0.005	0.10	0.43		26.95	3.959×10^{-7}		
0.0075	0.15	0.43	0.0025	33.00	7.417×10^{-7}	5.688×10^{-7}	4395
0.010	0.20	0.46	0.0025	40.77	1.428×10^{-6}	1.085×10^{-6}	2304
0.0125	0.25	0.48	0.0025	47.56	2.303×10^{-6}	1.866×10^{-6}	1339
0.015	0.30	0.502	0.0025	54.49	3.511×10^{-6}	2.907×10^{-6}	860
0.0175	0.35	0.508	0.0025	59.56	4.626×10^{-6}	4.069×10^{-6}	614
0.020	0.40	0.625	0.0025	78.33	1.082×10^{-5}	7.723×10^{-6}	434
0.0224	0.448	0.68	0.0024	90.00	1.674×10^{-5}	1.378×10^{-5}	181
						Total	10017

THE FRACTURE MECHANICS DESIGN PROCESS 159

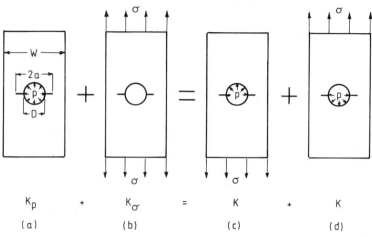

Fig. 8.10 *Superposition of configurations to obtain opening-mode stress intensity for pin-loaded lug (Example 8.5).*

Solution: Standard reference works do not contain Q-values for the appropriate configuration. A collocation solution is located for the correct geometry subjected to remote tensile loading, and also with internal pressure on the hole boundary [16] (Fig. 8.10(a) and (b)). By using the superposition illustrated in Fig. 8.10 it is possible to obtain the opening-mode stress intensity for the equivalent pin-loaded problem. The original and superimposed results are illustrated in Fig. 8.11. Note that for vertical equilibrium:

$$pD = W.$$

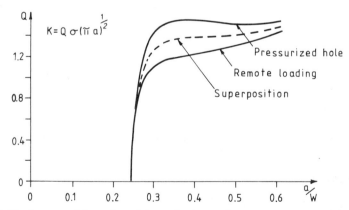

Fig. 8.11 *Superposition of results to obtain configuration correction factor for pin-loaded lug.*

Thus the appropriate pressure, p is given as:

$$p = W\sigma/D.$$

Since $2a/W = 0.4$, the Q-value may be extracted immediately as $Q = 1.38$. Thus, the stress intensity factor is obtained directly from:

$$K = 1.38\sigma(\pi a)^{1/2}.$$

The remote loading is composed of a constant load of 25 kN, and vehicle load of $0.45W$. Since the cross-sectional area is 1200 mm^2, and the semi-crack length $a = 12$ mm, the stress intensity at failure, K_{Ic} is given by:

$$K_{Ic} = 1.38 \left(\frac{0.25 + 0.45W}{0.0012} \right)(0.012\pi)^{1/2} \text{ MN m}^{-3/2}$$

Now K_{Ic} is a function of temperature, T, given by:

$$K_{Ic} = (0.2T + 70) \text{ MN m}^{-3/2}.$$

We may therefore express the limiting value of W which produces stress intensity K_{Ic} as a function of temperature in graphical form, as in Fig. 8.12.

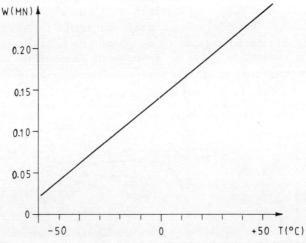

Fig. 8.12 *Maximum permissible vehicle weight (MN) versus temperature (°C), (Example 8.5).*

8.8 References

1. *ASME Boiler and Pressure Vessel Code* (1977).
2. *Airplane Damage Tolerance Requirements*, (1974), USAF Military Specification MIL-A-83444, 2 July.

THE FRACTURE MECHANICS DESIGN PROCESS

3. *Draft British Standard for the Design and Specification of Steel and Concrete Bridges* (1974), Part 10, British Standards Institution Document 74/13197 DC.
4. Fisher, J. W. (1974), *Guide to the 1974 AASHTO Fatigue Specifications*, American Iron and Steel Construction, New York.
5. Rolfe, S. T., Rhea, D. M. and Kuzmanovic, B. O. (1974), *Fracture Control Guidelines for Welded Steel Ship Hulls*, SSC-244, Washington DC.
6. Liebowitz, H. (Ed.) (1974), *Fracture Mechanics of Aircraft Structures*, AGARD-AG-176.
7. McGonnagle, W. J. (1971), 'Non-destructive testing' in *Fracture – An Advanced Treatise*, Vol. III, Ed. H. Liebowitz, Academic Press, New York.
8. *Acoustic Emission* (1972), ASTM, STP 505.
9. Rolfe, S. T. and Barsom, J. M. (1977), *Fracture and Fatigue Control in Structures*, Prentice Hall, New Jersey.
10. Rooke, D. P. and Cartwright, D. J. (1976), *Compendium of Stress Intensity Factors*, HMSO, London.
11. Tada, H., Paris, P. and Irwin, G. (1973), *The Stress Analysis of Cracks Handbook*, Del Research Corp., Hellertown, Pennsylvania.
12. Sih, G. C. (1973), *Handbook of Stress Intensity Factors for Researchers and Engineers*, Institute for Fracture and Solid Mechanics, Lehigh University, Bethlehem, Pennsylvania.
13. Gray, T. G. F. (1977), 'Convenient closed form stress intensity factors for common crack configurations', *Int. J. Frac.*, **13**, No. 1, 66–75.
14. Bowie, O. L. (1956), 'Analysis of an infinite plate containing radial cracks originating at the boundary of an internal circular hole', *J. Math & Phys.*, **35**, 60–71.
15. Tweed, J. and Rooke, D. P. (1976), 'The elastic problem for an infinite solid containing a circular hole with a pair of radial edge cracks of different lengths', *Int. J. Engng Sci.*, **14**, 925–33.
16. Parker, A. P. (1979), *Mechanics of Fracture and Fatigue in Some Common Structural Configurations*, RMCS Technical Note MAT/18.
17. Rooke, D. P., Baratta, F. I. and Cartwright, D. J. (1979), *Simple Methods of Determining Stress Intensity Factors*, to be produced as AGARD publication.
18. Tracy, P. G. (1979), *MMCLIB: A Computer Code for Stress Analysis Using the Modified Mapping Collocation Method*, AMMCR TR 79-13, US Army Materials and Mechanics Research Centre.
19. Pook, L. P. (1975), 'Analysis and application of fatigue crack growth data', *J. Strain Analysis,* **10**, No. 4, 242–50.
20. Matthews, W. T. (1974). 'Typical plane strain fracture toughness of aircraft materials' in *Fracture Mechanics of Aircraft Structures*, Ed. H. Liebowitz, AGARD-AG-176, pp. 509–79.
21. Lewis, G. I., Gardiner, R. W. and Peel, C. J. (1973), *Plane Strain Fracture*

Toughness Data for British Airframe Alloys, RAE TR 72224, Royal Aircraft Establishment, Farnborough.
22. *Damage Tolerant Design Handbook* (1972), MCIC-HB-01 Metals and Ceramics Information Centre, Battelle, Columbus, Ohio.
23. Hudson, C. M. and Seward, S. K. (1978), 'A compendium of sources of fracture toughness and fatigue—crack growth data for metallic alloys', *Int. J. Frac.*, **14**, R151–84.
24. *Stress and Strength* (1979), Vol. 7, *Crack Propagation*, Engineering Sciences Data Unit.
25. *Fatigue* (1979), Vol. 4, *Crack Propagation in Sheet*, Engineering Sciences Data Unit.
26. Brown, W. F. and Srawley, J. E. (1966), *Plane Strain Crack Toughness Testing of High Strength Metallic Alloys*, ASTM STP 410.
27. Rooke, D. P. and Tweed, J. (1973), 'The stress intensity factor of an edge crack in a finite rotating elastic disc', *Int. J. Engng Sci.*, **11**, 279–83.
28. Parker, A. P. and Farrow, J. R. (1981), *Stress Intensity Factors for Multiple Radial Cracks Emanating from the Bore of an Autofrettaged or Thermally Stressed, Thick Cylinder, Engng Frac. Mech.*, **14**, 237–41.

Index

Acoustic emission, 139
Aircraft construction, 140
Airy stress function, 1, 10
    in terms of complex stress functions, 11
Alternative stress intensity, 123 ff.
Analytic functions, 11
Analytical solutions, 49
Angled crack extension, 91
Arresters, crack, 140
Autofrettage, gun barrel, 151

Bend specimen, 113
Boundary collocation, 63 ff.
    of complex stress functions, 63
    convergence, 63
    formulation, 63
    geometry, special cases, 66
    outline programming technique, 67
    supplementary compatibility condition, 65
    symmetry properties, 66
Body force methods, *see* Boundary methods
Boundary elements, *see* Boundary methods
Boundary integral equations, *see* Boundary methods
Boundary methods, 74
    advantages, 75
    technique, 75
Boussinesq functions, 10, 15
Bridge structure, example, 157
Bueckner, H. F., Weight functions, *see* Weight function techniques

Cauchy–Riemann equations, 11
Clip gauge, 116
Closure, crack, 132, 149
Collinear cracks, example, 43
Collocation, *see* Boundary collocation and Mapping collocation
Compact tension specimen (CTS), 114
Compatibility of strains, 8, 9
    supplementary condition, 65
Compounding method, 76, 142
    accuracy, 77
    example, 77
Complex number, z, 11
Complex conjugate, $\bar{z}$, 11
Complex stress functions, 1, 11
    displacements in terms of, 12
    stresses in terms of, 12
    worked example, 13
Compliance, 25
    applied to finite element methods, 71
Compressive loading, cyclic, 146
Configuration correction factor, Q, 44 ff.
Conformal mapping, 15 ff.
    example, 17
Crack arresters, 140
Crack closure, 132, 149
Crack growth rate in fatigue, 123 ff.
    empirical, 124
Crack opening displacement, *see* Crack tip opening displacement
Crack path, 91
    stability, 97
Crack propagation rate, *see* Fatigue

Crack propagation (stable), 112
Crack shape, 33 ff.
  limiting values, 34
  reconstruction from weight function, 55
  used in derivation of weight function, 52
  with plasticity, 105
Crack tip opening displacement (COD), 103, 117
  plasticity, 101 ff.
  plastic zone, see Plastic zone
  stress field, 30, 36

Damage tolerance, 139
Determination of stress intensity factors, 49 ff.
Data location, 141, 142
Design using fracture mechanics criteria, 138 ff.
  examples, 142 ff.
Detection of cracks, 138 ff.
Displacements in cracked bodies, 28 ff.
Displacements in terms of complex stress functions, 12
Dugdale, D. S., Plastic zone model, 103
  crack shape, 105
  weight function solution, 104
Dye penetrant inspection, 138

Eddy current inspection, 139
Effective crack length, 102
Effective stress intensity factor, 89, 137
Elliptical hole
  degeneration into crack, 19
  stresses around, 17 ff.
Empirical crack growth rate results, 124
Energy considerations, 21
Energy, potential, 22
Energy release, 23 ff., 34

Energy release rate, 24, 25
  under fixed load conditions, 26
Energy, surface, 24
Energy, total, 24
Engineering shear strain, 7
Equilibrium, equations of, 4
  under plane stress, 10
Experimental methods, 78

Fail-safe design concept, 140 ff.
Failure criteria with moderate plasticity, 117 ff.
Failure assessment diagram, 120
Fatigue crack growth, 123 ff.
  and safe life design, 140
  effect of mixed mode loading, 135
  effect of overload, 133
  effect of R value, 132
  location of data, 142
  Paris, P. C., Crack growth law, 126 ff.
  pre-cracking of test specimens, 115
  threshold, 126
  variable amplitude loading, 133
Finite element methods, 70 ff., 142
  classical solutions, polynomial functions, 72
  compliance, 71
  isoparametric representation, 73
  J (or line) integral, 71
  non singular crack tip representation, 70
  singular elements, 72
Fixed grip conditions, 26
Fixed load conditions, 26
  energy release rate, 26
Flaw sizes, initial, 139
Force over an arc, 13
Fracture toughness, 31 ff.
  data location, 142
  example involving temperature variation, 157
  plane strain value, K_{Ic}, 31, 113
  testing, 113

INDEX

G, see Strain energy release rate
Gamma ray inspection, 139
General yield loads, 120
Geometry, special cases in collocation, 66
Green's functions, 51 ff., 142
Griffith, A. A., Crack, 21 ff.
Griffith, A. A., Solution for energy release, 23, 35
Gun barrel design example, 151

Harmonic operator, 10
Hooke's law, 8
Hole
    crack emanating from, 141
    mandrel enlargement of, 157

Irwin, G. R., Plastic zone model, 101 ff.
Initial flaw sizes, 139
Inspection intervals, 141
Inspection techniques, 138
Integral equation, 73 ff.
Integral transforms, 73
Integration of crack growth law, 126
    examples, 127, 130

J (or line) integral, 118 ff.
    in finite element methods, 71
    relationship to G and K, 118
'Just missed' crack length, 139

K, K_I, K_{II}, K_{III}, see Stress intensity factor and Modes of crack tip deformation
K_{Ic}, see Fracture toughness

Leak before break, 153
Life improvement, 145, 157
Line (or J) integral, 71, 118 ff.
Load–displacement record, 116
Loading conditions, 26

Magnetic anomalies detection, 139
Mandrel enlargement, 157
Mapping collocation, 68 ff., 142

Mapping techniques, 15 ff.
    example, 17
Mathematical shear strain, 7
Mises, see Von Mises
Mixed mode fracture mechanics, 89 ff.
    crack direction, 91
    fatigue crack growth, 135
    fracture conditions, 89
Mohr's circle, 5
Modes of crack tip deformation, 35 ff.
Mode I deformation, 35, 36, 91
Mode II deformation, 35, 36, 91
Mode III deformation, 35, 36, 91
Modified mapping collocation, 68 ff.
    limitations, 69
    stitching procedure, 69
Multiple load paths, 140
Non-destructive testing (NDT), 138 ff.
Numerical methods, see Boundary collocation, Boundary methods, Finite element methods, Integral equations, Mapping collocation and Modified mapping collocation

Opening mode, see Mode I deformation
Overlapping of crack surfaces, example, 61
Overload, effect on fatigue crack growth, 133

Pagoda roof method, see Rainflow method
Paris, P. C., Fatigue crack growth law, 124
    use of, 126
    examples, 127, 130
Penetrant inspection, 138
Physical law (Hooke's law), 8
Pin-loaded lug, 142, 157
Pipelines, 157
Plane deformation, 8
    stress, 9
    strain, 9

Plane strain fracture toughness, K_{Ic}, see Fracture toughness
Plane strain, plastic zone, 106 ff.
Plane strain, stress state, 110
Plane stress, plastic zone, 106 ff.
Plane stress, stress state, 110
Plastic deformation and acoustic emission, 139
Plasticity, failure criteria, 117 ff.
Plastic zone correction to stress intensity, 101, 103
Plastic zone radius, 102
Plastic zone shapes, 106 ff.
    effect of strain hardening, 107
    through the thickness effects, 108 ff.
Plastic zone, reversed in fatigue, 124
Polar coordinates, 3, 30
Potential energy, 22
Pre-cracking of test specimens, 115
Principal planes, 6
Principal stresses, 6

Q values, see Configuration correction factor

R curves (resistance curves), 111
R ratio (values), 132, 147
Radiographic inspection, 139
Radius of curvature at root of ellipse, 19
Rainflow method, 134
Range of stress intensity, 123 ff.
Residual stresses, 142
    crack closure, 133
    examples, 145, 151
Retardation due to overload, 133
Reversed yielding in fatigue, 124
Rice, J. R., see J integral
Riveted structures, 140
    approximate solution from weight function, 59
Rotor, example, 149

Safe life design concept, 140 ff.
Service life prediction, 141
Shear strain, 7
Shear stress, 2
Shot-peening, 157
Ship hull design, 140
Single edge notch test pieces, 113
Sliding (or shear) mode, see Mode II deformation
Sources of reference, 38, 142
Stability of crack path, 97 ff.
Stiffened sheet, 140
Strain, 6 ff.
    direct, 6
    shear, 7
Strain hardening, effect on plastic zone, 107
Strain energy density, 93
Strain energy density criterion, 93 ff.
    physical significance, 95
Strain energy release rate, G, 24 ff., 38, 39
    in derivation of weight function, 53
Stress
    around elliptical hole, 17 ff.
    definition, 1
    near crack tips, 36 ff.
    notation, 2
    principal, 6
    relief example, 145
    tranformation, 5
Stress function
    Airy, see Airy stress function
    Boussinesq, 10, 15
    complex see Complex stress functions
    for pressurized crack, 32
    for remotely loaded crack, 29, 31
    Westergaard, see Westergaard, H. M., Stress function
    Williams, see Williams, M. L., Stress function
Stresses in cracked bodies, 28 ff.

INDEX

Stress intensity factor, 30 ff.
  definition, 30
  derived from weight function, 53
  determination, 49 ff.
  in standard test pieces, 113 ff.
  in terms of G, 31
  locating solutions, 141
  range, 123 ff.
Stress wave emission, 139
Stitching applied to collocation, 69
Subcritical crack growth, 123 ff., 139
Suffix notation, 2, 7
Superposition principle, 31, 159
Surface energy, 24
Symmetry properties in collocation, 66

Tearing mode, *see* Mode III deformation
Temperature, example of effect on toughness, 157
Testing, plane strain fracture toughness, 113
Thermal inspection, 138
Threshold in fatigue crack propagation, 90, 126
Through the thickness effects, plastic zones, 108 ff.
Toughness, *see* Fracture toughness
Transfer of loading to crack line, 53
  example, 33
Transformation of stress, 5
Tresca yield criterion, 101
  plastic zone shapes, 106 ff.

Types of failures, 108

Ultrasonic inspection, 139
Uniaxial stress, crack under, 28 ff.
Uniqueness of the weight function, 53
  proof, 79

Variable amplitude loading in fatigue, 133
  rainflow method, 134
Visual inspection, 138
Von Mises' yield criterion, 101
  plastic zone shapes, 106 ff.

Weight function techniques, 51 ff., 141, 142
  body forces, 53
  examples, 54, 56, 60, 145
  mixed boundary conditions, 81
  uniqueness, 53, 79
Welded component, 145 ff.
  crack closure, 149
Westergaard, H. M., Stress function, 39 ff.
  examples, 40, 41, 43
Williams, M. L., Stress function, 44 ff.

X-ray inspection, 139

Yield criteria, 101
Yield loads, 120
Yielding around crack tips, 101 ff.
Yielding, life enhancement by, 157
  reversed, in fatigue, 124